T0073547

The Age of Prediction

The Age of Prediction

Algorithms, AI, and the Shifting Shadows of Risk

Igor Tulchinsky and Christopher E. Mason

The MIT Press
Cambridge, Massachusetts
London, England

© 2023 Igor Tulchinsky and Christopher E. Mason

All rights reserved. No part of this book may be reproduced in any form by any electronic or mechanical means (including photocopying, recording, or information storage and retrieval) without permission in writing from the publisher.

The MIT Press would like to thank the anonymous peer reviewers who provided comments on drafts of this book. The generous work of academic experts is essential for establishing the authority and quality of our publications. We acknowledge with gratitude the contributions of these otherwise uncredited readers.

This book was set in ITC Stone Serif Std and ITC Stone Sans Std by New Best-set Typesetters Ltd. Printed and bound in the United States of America.

Library of Congress Cataloging-in-Publication Data

Names: Tulchinsky, Igor, 1966- author. | Mason, Christopher E., author.
Title: The age of prediction : algorithms, AI, and the shifting shadows of risk / Igor Tulchinsky and Christopher E. Mason.
Description: Cambridge, Massachusetts : The MIT Press, [2023] | Includes bibliographical references and index.
Identifiers: LCCN 2022016359 (print) | LCCN 2022016360 (ebook) | ISBN 9780262047739 (hardcover) | ISBN 9780262373197 (epub) | ISBN 9780262373203 (pdf)
Subjects: LCSH: Artificial intelligence—Social aspects. | Predictive analytics—Social aspects.
Classification: LCC Q335 .T85 2023 (print) | LCC Q335 (ebook) | DDC 006.3—dc23/eng20230106
LC record available at https://lccn.loc.gov/2022016359
LC ebook record available at https://lccn.loc.gov/2022016360

10 9 8 7 6 5 4 3 2 1

Contents

Introduction

Humanity has entered a new era. We are now living in a world that is increasingly wired by billions of predictive algorithms, a world in which *almost everything* can be predicted and risk and uncertainty appear to be diminishing in almost all areas of life. We are living longer, thanks to advances in health care and precision medicine. We have a greater mastery of the physical world that allows us to dream of, and build, new technologies that allow us to explore other planets and visualize billions of galaxies. We can model markets, disease, and traffic with increasingly greater precision, and we're getting very close to handing over the keys to the car so it can drive itself. Even more striking, our tools may be revealing the genesis of some elusive and stubborn complexities of human behavior, and algorithms are even being used to alter people's behavior. Predictive algorithms have changed the world, and all the worlds to come, and there is no going back.

To better understand this new era, this book is focused on two abstractions that are increasingly shaping our lives: prediction and risk. Prediction speaks of what is to come; risk, less visibly, calculates the probability that a model is wrong and, like a grim accountant, totals up the costs of that error. Where prediction serves as a light projecting into the likely future, risk is the shadow of what cannot be seen or predicted. Their relationship is often paradoxical, particularly as prediction changes age-old apprehensions about risk and reshapes the very type of society and world in which we live.

We call this new era the Age of Prediction. In this emerging age, prediction is growing more powerful, ubiquitous, and precise, both because the tools for making predictions are advancing quickly and because the fuel of prediction—data—is accumulating at such a rapid and exponential (sometimes super-exponential) rate. Yet, none of this is easy. When it comes to many natural processes, we have grown more and more quantitative—and thus more predictive. Even quantum physics, a field that has seen some of the most dramatic scientific advances in the past century, requires the ability to precisely calculate elements of indeterminism, randomness, chaos, and probabilistic functions. Biology and medicine now require the same ability, as they adapt to ever-changing genetic mutations and environmental variables, such as drug-resistant tumors or an invasive species disrupting an ecosystem.

There are limits to our reach, of course. The COVID-19 pandemic emerged before we could stop it. But, as we will see in chapter 1, companies were rapidly able to develop vaccines against COVID-19 using predictive algorithms, and as chapter 5 explains, epidemiologists and geneticists have grown more adept at modeling infectious-disease spread once it begins. Modern medicine generally has become increasingly precise, personal, and predictive, especially when it comes to identifying and even repairing dangerous mutations.

However, the challenge of dealing with social processes is more difficult. Any trend fueled by the decisions of one or many human agents involves more risk—sometimes much more. As the economist John Kay and the former Bank of England governor Mervyn King note in their book *Radical Uncertainty* (2020), "The world of economics, business, and finance is non-stationary—it is not governed by scientific law." People are often the hardest things to predict, and that makes prediction more difficult in these social areas.

Regardless of this difficulty, we now have the tools and the data to predict more things with greater accuracy and farther out in time. In the Age of Prediction, this trend will continue, expand, and accelerate, and risk will shrink but not disappear. Risk is a realm of the unknown, of the incalculable and uncertain. The relationship between prediction and risk is conventionally an inverse one: if our ability to accurately predict the weather a week ahead were ever to approach 100 percent, the risk that you would fail to bring your umbrella on a rainy day would, in theory, approach zero (you might forget the umbrella or lose it, but the likelihood of this too could be quantified). Prediction in natural, stationary systems can approach perfection once those processes are truly understood; large numbers drive certainty so close to 100 percent that we can declare a process a scientific law. Empirical proof of this phenomenon, known as inverse probability, was scribbled down as a now-ubiquitous equation in an essay by the then obscure nineteenth-century British nonconformist cleric Thomas Bayes. From him, we have Bayesian models, in which new data continually update probabilistic judgments about future events.

Yet, the relationship between prediction and risk is not as simple as getting more data, running them through an algorithm, and applying them broadly. For instance, whose risk are we talking about? In chapter 8, we discuss the rise of autonomous, predictive weaponry. A military that unleashes such weaponry obviously believes it will be effective in military terms; smart weapons may kill more of the enemy and fewer of that military's own soldiers. However, given the competitive nature of armed conflict and the cross-fertilization of technologies such as artificial intelligence (AI), robotics, and modern munitions, any prediction about the evolution of warfare is profoundly uncertain and represents a rising risk.

There are other ways to look at prediction and risk. The convention that better predictions mean reduced risks may well reflect how the human condition has evolved since our emergence as self-aware, humanoid animals in Africa. The human brain has evolved into an instrument of forethought, planning, and prediction. One could argue this is a quintessential aspect of being human: projecting dreams, plans, and ideas into the future. For many millennia, prediction was more accurate under relatively simple conditions and over short periods of time—for planting, hunting, and fighting. Much of life remained uncertain, and much thought went into the role of what seemed like fate. Death appeared suddenly and arbitrarily. Disaster lurked at every turn. Risk was omnipresent.

The advent of modern science began to push back that darkness, at first very slowly. The seventeenth century brought us advances in physics and mathematics, including calculus, which enabled Isaac Newton to predict accurately the regularities of the heavens and the flight of bullets. That same period also saw the first breakthroughs in what we now know as probability and statistics. A better understanding of probability opened the door to a deeper understanding of random processes such as chance and luck, and statistics drove efforts to gather data and analyze them with sophistication. By the nineteenth century, data gathering and analysis had entered a capitalist world that focused increasingly on prediction; significantly, this period saw the development of robust futures markets, which consumed agricultural statistics and weather forecasts to set national and global prices. Later, but no less important, advances came from a deeper understanding of life itself, from evolution to genetics to genomics, unleashing the powers of modern medicine. All these advances have altered myriad perspectives on risk, spanning predictions of a single cell's response to a drug or a single stock's fluctuations or an individual's likely next vote. As we describe in this book, a surprising common thread of prediction is that the methods used in one discipline can inform another; metrics like the Gini coefficient can measure economic differences, yet we can also use the same formula to map shifts in bacterial DNA to predict growth and resistance to antibiotics.

Predictive algorithms have fundamentally changed almost all aspects of our world, yet such a change has been brewing for several centuries. In her book describing the changing relationship between prediction and uncertainty in late nineteenth-century America, the historian Jamie Pietruska offers up striking conclusions about that era. Americans quickly grew accustomed to predictions in many areas of life, but they also grew skeptical of those predictions' accuracy and more willing and able to discriminate between scientific efforts and fortune-telling or phrenology, as well as to accommodate the potential corruption stemming from the rising value of data and prediction. Yet even as new predictive technologies surfaced, new risks appeared: economic instability, social dislocation, and inequality as well as, Pietruska notes, "industrial accidents, steamboat explosions, and railroad collisions." Americans changed their behavior in nuanced ways as they struggled with what she calls "the spectre of Uncertainty."

In our book, we are focused on examining changes in behavior that are likely to arise as prediction grows better and risk changes. We look at the history of predictive tools, the current armamentarium, and the likely future growth and application of these algorithms and powerful tools. We ask the following questions: If prediction in, say, forensics, genomics, markets, warfare, politics, and other arenas grows significantly better, how will that change the apprehension of risk and the resulting human behavior? How will the world be better or worse or both?

Behavioral responses to changing risk have undoubtedly always been fluid, but at a rate of change that was barely discernible. Now we are beginning to see more rapid spikes in predictive capacity and more challenging responses. Many of these responses are counterintuitive or paradoxical, resembling Pietruska's description of a Gilded Age of both greater predictability and greater uncertainty. The risk comes in different guises. For instance, insurance has employed the term *moral hazard* to describe a common risk undertaken by an insurance provider since the mid-nineteenth century. At first, it referred to a kind of customer who was viewed, often for biased reasons, as a bad risk: a minority, a woman, or a foreigner. In time, moral hazard grew more focused on incentives that increase rather than reduce risk. Yet, holders of insurance may have less incentive to behave prudently—the classic example is the entrepreneur who insures his failing business before torching it—thus accentuating risk. Today, the term, as often used by economists as by insurers, has taken on an even broader, less moralistic or ethical meaning. Moral hazard has more recently been attributed to financial players who shoulder large risk while believing that, if things go badly, they will be bailed out. Dangerous incentives can occur when predictability increases and risk appears to decline or disappear. Predictably, when individual risk approaches zero, behavior can grow more reckless.

Risk can also take on unexpected characteristics produced by feedback mechanisms. As you reduce risk in some systems, you may also drive new complexity, thus producing new opportunities for risk. If risk depends on who is engaging with it, it also depends on how that person is thinking about it. That is, risk can be subjective—more so individually than collectively. This makes sense when you realize that purely rational decision-making, the foundation of much economic theory, remains elusive.

One of the most significant outgrowths of predictive technologies is the demand for the kind of data that many people view as intimate or private. This issue, of course, is increasingly contentious. Early genetic analysis led to the first legislation regulating who gets access to the results and what they can do with that data. But the trend continues to spread: privacy issues show up in many of the following chapters in areas ranging from genomics to forensics to politics. Even today, the feedback loop created by individuals reacting to demands for intimate data or real-time monitoring has spawned a wide range of new risks, which may include people not only refusing to share their own data but also rejecting a hold on the very realities that underlie that risk. Ironically, this can limit predictions and increase risk, but still satisfy a need for some to feel they have more control.

As coauthors, we come from different fields—finance and medicine—but much of the way we view the world has been shaped by our efforts to make predictions and to reduce risk in our separate, yet sometimes overlapping, careers. The chapters in this book focus on areas we know quite well, but we also apply the lessons from our own fields to emerging areas to test some ideas about prediction and risk. These ideas are, as is this book, speculative in nature. We are imagining a future where the ground is shifting and the "conventions," as described by the Scottish philosopher David Hume, are changing.

As the dawn of the Age of Prediction rises, there are clearly some dangers, but we believe that over time the greater scientific ability to predict will prove broadly beneficial. We are optimists. But we also recognize the challenges and the unknowns that enhanced prediction carries with it. Predicting human behavior is a particularly difficult task. We realize that predictions about prediction are a risk in themselves, and that the light from our data and our algorithms reaches only so far into the future. Nonetheless, we cannot escape the powerful currents sweeping humans toward greater and greater predictability. Only by exploring its complexities, across many disciplines, can we begin to master both prediction and the changing nature of risk.

1   Prediction and Risk

Virology is not a field most people have mastered. At some point, everyone must cope with a virus—the flu, a cold, measles—but even the distinction between a virus and a bacterium is a nuance lost on many. As a result, 64 percent of people interviewed in a World Health Organization study believed that viruses could be knocked out by antibiotics, which is not only false but also deepens the problem of bacterial resistance to antibiotics due to overuse. In fact, viruses are far smaller than bacteria, 20–500 nanometers versus some bacteria that are 100 times larger.

Viruses can be seen only through an electron microscope and are relatively rudimentary: anywhere from four to several hundred genes wrapped in a protein coat. They are cellular parasites that can reproduce only by invading a cell and commandeering its genetic machinery to make copies of itself. There are two broad categories of these tiny molecular machines, DNA viruses and RNA viruses, and the latter mutate particularly quickly. That mutational instability makes them adept survivors. An RNA virus such as Ebola is structurally simple—a thin, wormlike filament, or *filovirus*, containing only four genes—yet is savage in its attack and elusive in its pattern of emergence, remission, and reemergence. Researchers have spent decades trying to find the animal reservoir for Ebola. Some think it might be a fruit bat or primates, but the quest is ongoing to find the hosts of Ebola as well as of many other viruses that can come from animals, known as zoonotic diseases.

We can't escape viruses, which are ubiquitous, but our immune system disposes of most of them painlessly and quickly. Vaccines can provide protection from viruses that can't be naturally or easily repelled, such as smallpox and polio. However, the real threat posed by a virus occurs when it becomes unrecognizable by an individual's immune system, especially if the virus has mutated in a way that makes it more infectious and more lethal. That mutational process is, like all evolution, a blend of random chance, drift, and selection, and thus difficult to predict.

The challenge of prediction is perhaps nowhere more evident than in the COVID-19 outbreak that started in December 2019. The outbreak told us about our limited ability to more accurately *predict* complex change as well as about the social and economic challenges of responding to changing perceptions of *risk*.

The chance of a novel virus was well known at the end of 2019, but most people and politicians viewed its spread as an unlikely event. Since the 1960s, a parade of malevolent viruses had caused dreaded diseases such as Ebola, AIDS (acquired immunodeficiency syndrome), SARS (severe acute respiratory syndrome), MERS (Middle East respiratory syndrome), Zika, swine flu, and bird flu, but people were mollified by the fact that the outbreaks were quickly contained. Yet infectious-disease experts and scientists predicted that more were coming. Movies such as *Contagion* (2011) and fiction best sellers such as Richard Preston's *The Hot Zone* (1994), about a grisly Ebola outbreak, were built around the notion of modern plagues, as were more sober, scientific books, such as Laurie Garrett's pioneering work *The Coming Plague* in 1995 and David Quammen's *Spillover* in 2013. In his book, Quammen noted that more than 30 million humans had died since 1981 from "the emergence of virulent new realities on this planet"—he specifically warned of *zoonosis*, an infectious disease that can be transmitted from various animals to humans—and notably pointed to the chances of a dangerous virus coming from China. Even the US government was aware of the potential for such a pandemic. In 2019, the US Department of Health and Human Services engaged in a series of mock exercises called "Crimson Contagion" based on a fictional new virus brought to the United States by a single American tourist returning from China and attending a musical event in Chicago. The exercise posited some 586,000 US deaths in this scenario.

Yet when the SARS-CoV-2 pathogen (the virus that causes COVID-19) arrived in the United States, few were prepared. Like a financial crisis, the often-discussed pandemic ironically still came as a surprise to almost everyone, and it proved to have systemic repercussions, rippling through areas that seemed to have no relation to public health: state and federal relations, racial and class conflicts, supply chains, education, technology, mental health, entertainment, and myriad social and economic strife.

The first official reports of COVID-19 came from Wuhan, China, in December 2019, with most evidence pointing to the first cases clustered around a "wet market," where livestock, fish, and other animals are bought, transferred, and moved. But in the United States, reports about the new virus did not emerge until early January 2020, and most people presumed it would stay in Asia. After all, China had experienced other novel zoonotic virus outbreaks, but their effects on the United States, although newsworthy, were ultimately minor. The terror of the global flu epidemic of 1918–1919 was an exception—the disease killed some 50 million people around the world—but that happened a century ago, and most people assumed they were less susceptible in 2020. The bubonic plague was another epidemic, but it struck medieval Europe; the global hubris in 2020 presumed that humanity would be less vulnerable. However, all the preceding pandemics had existed in a time without rapid, global mass transit. Whereas older viruses had to move by foot, boat, or train, SARS-CoV-2 spread as quickly as a plane could fly. The first cases of community spread in the United States emerged in Washington State, then in New Rochelle, 20 miles north of New York City. Quickly cities, regions, and whole countries fell into lockdown like dominoes. By early March 2020, Italy had shut down, then California, then New York. By the middle of that month, much of the developed world was struggling to cope with the invisible, still poorly understood virus.

What ensued was a series of unprecedented—and by definition surprising —consequences. Schools closed. Offices shuttered. Travel stopped. Restaurants, bars, sports, hair salons, and bowling alleys shut down. Hospitals in hard-hit areas were overwhelmed, and COVID-19 deaths swept through nursing homes, prisons, and industrial facilities such as meatpacking plants. The federal response was at best uncertain, at worst dysfunctional: states scrambled to grab scarce ventilators and buy basic protective equipment such as masks, gloves, gowns, and swabs; testing was sporadic, and the reagents needed for testing were sparse; data were problematic. Shortages in toilet paper and meat ebbed and flowed. The US economy didn't die, but at least a quarter of it went idle, and the unemployment rate jumped to 14.7 percent, the highest level since the Great Depression.

No one really had a clue about what was coming next. Millions of workers were laid off, particularly in smaller businesses that had to close; some

22 million signed up for unemployment support in April. Economic growth (as measured by gross domestic product, or GDP) dropped by 32.9 percent from April to June 2020. Prices of commodities, notably oil, collapsed as growth slowed. The price of West Texas Intermediate crude oil dropped to negative levels for the first time in history as producers ran out of places to store excess oil and had to pay others to dispose of it. Financial markets, perhaps the most influential of predictive-fever charts, began to tumble, reviving memories of the collapse of Lehman Brothers in 2008. Stocks in the S&P 500 peaked on February 16; as news of the virus spread, the index began to slide, and by March 24 it had fallen some 34 percent. With lockdowns spreading, particularly on the East and West Coasts, commentators darkly declared the possibility of another Great Depression.

Then, just as suddenly as COVID-19 had arrived, stocks bottomed out, reversed, and soared; on April 6 the S&P 500 rose 7 percent. There was knowingly gloomy talk of the proverbial "dead-cat bounce." The narrative that the world was heading into a long-lasting depression persisted after the recession was officially declared over. Many remained deeply skeptical of the continuing rally, which was widely viewed as an example of how Wall Street is rigged for the wealthy.

Wall Street had its own explanations for the reversal, though. In late March, the Federal Reserve had taken its first aggressive steps to lower interest rates to near zero, promising to keep them there for some time and to buy billions of dollars in securities every month, thus providing liquidity to the financial system. The Fed also ramped up direct lending to banks, securities firms, and corporations. Congress, shocked into bipartisanship, passed the first of three major COVID-19 aid packages. On the health side, there were signs—sketchy but hopeful—that the number of cases in the hardest-hit areas of the Northeast, in particular New York City, was plateauing. The draconian lockdown seemed to be working. Dozens of vaccines were heading to clinical testing. At the same time, however, conspiracy theories, soon to be associated with both the origins of the virus and the pandemic response, emerged.

Those were the headline events. But there were other, quieter signs of an upturn. Quantitative investment firms across Wall Street had built literally millions of algorithms to exploit marketplace signals of all kinds. *Algorithms* are computer code, instructions that send an input through a finite number of steps to produce some sort of output. Today, they are ubiquitous, although as invisible as the COVID-19 virus itself, and, like the growth of data and computer power, they continue to proliferate exponentially.

Algorithms are predictive models of data. A financial algorithm attempts to predict the movement of a stock, bond, commodity, or derivative price; macro- and microeconomic data; intrinsic values; or the state of dynamic relationships between financial instruments and data. An algorithm can establish either a correlation between two data points (if one increases, another rises in tandem), an inverse correlation (one data point falls, driving up another point, thus providing a hedge), or other nonlinear relationships. Algorithms can produce strong or weak signals or die out entirely. In their totality, algorithms resemble a vast array of sensors responsive to a diverse and profoundly complex flowing stream of signals.

By late March 2020, many of the financial algorithms that had ceased to function in the sudden collapse brought on by the pandemic just as suddenly began to show signs of life. This reversal came days before the market itself began to rally, and the outlook was still bleak. The stock markets were reeling with panic, massive selling, and huge losses. All kinds of organizations, including hedge funds, had liquidated their holdings; cash was the only safe haven. Analysts slashed corporate-earnings estimates, particularly for hard-hit industries: airlines, cruise lines, hotels, commercial real estate. City dwellers fled to the suburbs or beyond, like the characters in Boccaccio's *Decameron* gathering in the countryside to escape the Black Death.

But in the midst of this free fall, those predictive algorithms suddenly began suggesting not only that the market was functioning but also that enormous opportunities were dawning and that the time to act was *now*. The algorithms were saying it was time to buy again. And, in fact, investors began to return to the market, and stocks stabilized, then moved upward, less concerned with the grim state of the present and more focused on a postpandemic future.

The movement back to the market was a bet that the future would be better than the present. It was driven by the belief that the COVID-19 virus would be defeated or would burn itself out, which to many people, still stuck at home mastering sourdough bread, sitting through Zoom calls, and schooling their kids, seemed to be an immensely long shot. By

December 2021, though, some 20 months later, the rally was still rolling along, even though 2022 would prove to be more problematic for the markets for a number of reasons, some pandemic-related, others entirely independent.

Predicting a Vaccine

Humanity needed a vaccine to protect itself against the virus, but, as with the markets, pessimism hung over the possibility that such a panacea would develop quickly. Vaccine development had become a tough business. The technical challenges were large; the regulatory approvals came slowly, given the reality that millions of people would receive shots, and the business itself wasn't a huge spinner of profits. In addition, a small (but noisy) minority of critics thought vaccines were dangerous. Traditional vaccines—which use an inactivated virus, or components of a virus, to induce immunity— took as long as a decade to develop and test. Many never made it to market. Vaccines as a commercial product suffered from a "productivity gap": the expected revenue from new vaccines wasn't high enough to justify the effort and required investment. These factors forced the industry to consolidate, thus reducing competition and driving up prices.

But COVID-19 forced a reconsideration of the economics of vaccine development, which now became a huge potential market with significant and immediate global need. Fortuitously, years of investment had created innovative tools that allowed vaccine development to move much, much more quickly and much more effectively.

On Saturday, January 25, 2020, in Mainz, Germany, an ancient city on the Rhine River, those factors met. At a company called BioNTech, its husband-and-wife cofounders, Ugur Sahin and Özlem Türeci, read an article in the British medical journal the *Lancet* about the outbreak in Wuhan. They thought they might be able to develop a vaccine quickly. There wasn't a lot of bureaucracy to deal with: the pair, children of Turks who had settled in Germany, had founded BioNTech in 2008 and taken it public in September 2019. For a decade, they had researched a new approach in vaccines, known as messenger RNA, or mRNA, for use first against cancer and then against infectious diseases. Before the weekend was over, they had used a suite of predictive algorithms and the newly sequenced genetic code of the virus, first released by Chinese researchers only two weeks earlier, to craft ten vaccine models for COVID-19; ten more would follow in the weeks ahead.

BioNTech's vaccine used a coded sequence of mRNA—the single-stranded nucleic acid that transmits instructions from DNA to the cell's ribosomes, which assemble proteins—to induce the production of SARS-CoV-2 viral proteins, and eventually antibodies against the virus. The company used sophisticated sequence analysis and software that predicted how the proteins would fold. Specifically, its models homed in on RNA coding for protein components of the virus that might be most capable of triggering a strong immune response while also being stable enough for mass production. This RNA strategy broke with traditional vaccine development, which used attenuated or dead viruses to induce immunity. An mRNA vaccine was relatively easy to design and manufacture and was exquisitely targeted to the most important part of the virus, the spike protein, which studs the outer membrane of the virus and is used to penetrate human cells. In contrast to a traditional vaccine, in which a body's cells are ravaged as the immune system tries to learn all 30,000 letters of the genetic code of a virus the size of SARS-CoV-2, an mRNA vaccine means the immune system has to master only 4,284 letters. (BioNTech was, of course, not alone in the search for a vaccine. Scientists at the US biotech firm Moderna quickly followed, building models for its own mRNA vaccine.)

In a few weeks, BioNTech and the US drug giant Pfizer, which had already been partnering on mRNA applications for infectious diseases, planned and then launched human clinical trials; in three months, more than 120,000 people were enrolled. In nine months, the vaccine had gone from a prediction on a screen to 95 percent efficacy and received an emergency-use authorization from the US Food and Drug Administration, and eventually full approval.

BioNTech's use of predictive algorithms in vaccine development was hardly unique, but the use of machine learning, a form of AI in which the computer learns from the data over time, identifies patterns, and makes decisions, was integral to the explosion of vaccine work that has come from the COVID-19 pandemic. During their computational design of the vaccine, scientific teams could compare the sequence data from all other known viruses and tests. Before the end of 2020, the World Health Organization counted 34 vaccine candidates being tested in humans and an additional 145 tested on animals or in the lab. Predictive algorithms were used to understand viruses better, predict what design elements would generate the greatest immune response, track the evolution of variants, and make sense of both experimental and clinical trial data.

Stocks and markets have been around since the early seventeenth century and vaccines since the eighteenth century, yet never before have the algorithms used in drug and vaccine discovery and the algorithms that shape much of the market had so much in common. These algorithms, fed by oceanic amounts of data and increasing computer power, define the Age of Prediction.

2   The Complexity of Prediction

No one, of course, would say that the COVID-19 pandemic demonstrated a triumph of prediction. As then–New York governor Andrew Cuomo often declared in his much-watched daily briefings during the initial COVID-19 surge, public-health officials lagged behind the virus at every turn. Not only during periods when infections exploded exponentially, but also in the secondary and tertiary consequences of the reopening, we saw the snarl of supply chains, the inflationary surge, the disparity in deaths, and even the misinformation, antivaccine and antimask sentiment, and conspiracy mongering that emerged. The Crimson Contagion report was wrong in one fairly significant way: it had *underestimated* the US death toll.

Yet, COVID-19 also showed the speed and power of the algorithms that drove a variety of predictive technologies. Such algorithms could be used for more than just stocks and vaccines. Corporations increasingly applied predictive technologies, often using machine learning, to a variety of business challenges that arose throughout the pandemic: how to handle inventory, what to stock, when to buy, how to understand the changing nature of consumer demand. The role of predictive technologies in the COVID-19 pandemic was a classic example of a counterfactual. The virus presented nearly every individual and every institution with a radically altered reality. The question is not really whether this pandemic was a failure of prediction but, rather, how bad it would have been if a variety of predictive technologies, often operating behind the scenes, had not been in place. Would markets have responded so quickly? Would vaccine development and testing have moved so rapidly from cell culture to animal and human testing and then to the arms of billions of people? Also, will some of these postpandemic effects—chip shortages; supply-chain snarls; inflation in commodities, energy, and food—prove to be more easily remediable because of the predictive algorithms that currently exist? So far, yes.

Today, prediction can ascribe features and probabilities to each cell of an embryo, to each biochemical change in an astronaut in space, to each cancer patient, to each tendency in financial markets, to complex natural processes, and to social behavior. A key word to describe this ability is *complex*. Prediction in complex nonlinear systems has traditionally been nearly impossible because these systems often appear to behave randomly: think about trying to predict the course of a turbulent stream of water, consumer choice in an economy, the actions of an angry mob, or the mutations that drive evolution.

In the past, both the tools and the computational power to penetrate such complexity and to determine an underlying order were absent. Markets seemed to take a "random walk," and genetic mutations appeared to be arbitrary copying errors and a mostly opaque selection process. Earthquakes, volcanoes, the weather, climate change, schools of fish, flocks of birds, crowds, and crime waves react to nonlinear inputs that until recently have been beyond our ability to recognize, understand, and anticipate.

So, what has changed? Three things: the accelerating growth in computing power—driven mostly by Moore's law, which describes the doubling of capacity for every generation of semiconductors (roughly every 18 months); the explosively exponential growth in data (especially genomic data); and the development of increasingly sophisticated analytical processes using machine learning and AI. The algorithms can now model all of the data to continually improve the predictions, and it doesn't matter which aspect of the data or metadata creates the improved model. All data is fuel for the prediction.

In time, the powers of prediction will be central to almost all our activities and products. Today, we can already see the broad outlines of this centrality, and we know that it will spawn consequences that few have anticipated or considered in depth: certain kinds of risk will decline, but, paradoxically, that decline will be one of the major challenges in a world of advancing predictability.

The Role of Moral Hazard

Normally, prediction and risk are inversely correlated. Since an upright, apelike animal resembling *Homo sapiens* emerged on the scene millions of years ago, the relation between these two factors has consisted of weak prediction and overpowering risk; one theory of the nature of human consciousness is that it evolved to cope with that disparity, to constantly update inferences based on sketchy and changeable data. Humans are, at least in part, Bayesian calculators. With a few exceptions, such as the seasons, humanity could not be confident of anything beyond the next few hours or days, and even that span could be a stretch. As a result, humankind has long been fixated on the inexplicable determinations of fate, which the ancient Greeks depicted as three goddesses, the Fates, who determined one's path in life: Clotho spun the thread of life, Lachesis dispensed it, and Atropos cut it, signifying death. Escape from life's manifold hazards came only outside space and time: in heaven, a world of ideal forms, or through the second sight of an officially sanctioned oracle, such as the one at Delphi. That sense of life as a plaything of the gods or God—as in Jonathan Edwards's eighteenth-century sermon "Sinners in the Hands of an Angry God"—began to slowly change only with the Enlightenment as mathematicians and physicists explored increasingly complex aspects of nature.

As Peter Bernstein, an investment manager, noted in *Against the Gods: The Remarkable Story of Risk* (1996), "The revolutionary idea that defines the boundary between modern times and the past is the mastery of risk: the notion that the future is more than a whim of the gods and that men and women are not passive before nature."

Bernstein's history features a slow and steady accumulation of insights and knowledge over many centuries, but that slow accumulation begins to accelerate in the twentieth century, launching an era of exponential growth. In 1996, the human genome hadn't yet been elucidated; that would take until 2001. Artificial intelligence as a field was struggling to attract funding. The internet was still in its infancy, and the smartphone was science fiction. Few could really anticipate the explosive growth in data that was coming. Bernstein even touched on the emerging development of neural networks, genetic algorithms, chaos theory, and the belief "that all complex systems like democracy, the path of technological development, and the stock market share common patterns and responses." But he was skeptical that these techniques could truly be predictive. "Nothing is more soothing or more persuasive than the computer screen, with its imposing arrays of numbers," he wrote, noting, "Likeness to truth is not the same thing as truth." Anticipating surprise—what he called "wildness" in nature, markets, or human conflict—would remain difficult, if not impossible.

What Bernstein could not see is how we would be swept forward by a tsunami of data and computing power and how that immense growth would begin to hone our predictive tools and make them more effective, and even a little eerie, like when algorithms (e.g., ChatGPT) can create almost perfect Shakespearean prose. As models improve, as the tools grow faster and become less expensive, and as the data flows are larger and more diverse, it is tempting to imagine that we can eliminate much of the variance in prediction and risk, particularly as we improve the iterative learning process at the heart of machine learning.

For example, if we can monitor genetic alterations that occur over time, we can predict health problems a year, a decade, or even a quarter century early for patients. We can also get a jump on infections before they become epidemics, guess the likely reaction of cancer cells to targeted drugs, and prepare for problems such as antibiotic resistance, which may kill millions of people per year by 2050, according to a study by Marlieke de Kraker and colleagues from 2016. Personal and predictive medicine will have its analogue in personal and predictive insurance. AI will improve everything from weather forecasting to earthquake prediction to disease diagnosis—indeed, it already has in some ways.

However, none of this will occur easily or without disruption. An autonomous vehicle is predictive, but it will never become ubiquitous if it doesn't significantly reduce the risk that it will hit something. Sales and marketing are less risky but more subjective. They have already attempted to apply predictive algorithms to our tastes and desires, with promising if mixed results. Forensics is being transformed—the ability to trace DNA left behind by killers has already cleared up thousands of cold cases—but predictive enterprises such as hiring and polling remain little changed.

This is a book about the next few decades, which we can already see quite clearly, when improved prediction will reduce risk, but won't sweep it away. Aging, mortality, and the hazards of everyday life will remain. Also, we know that these changes in risk will change how people and society act and react.

One embedded trade-off of greater certainty is the human propensity to take on more risk. Here again we see moral hazard, a familiar concept for insurance companies and financial regulators. When people are convinced that their risk is much reduced, they tend to engage in more or different activities, climbing out on the risk curve, whether they're building homes on floodplains or betting their 401(k) accounts on dot-com stocks. Bernstein called this the "seat-belt problem." Seat belts convince people that they are safer; one result of their use is fewer deaths on the road, but another is more reckless driving and more accidents. Particularly in markets, risk tends to be correlated with reward. How does the reduction of risk alter this well-known relationship? How will tools of prediction remake and disrupt entire industries—and nations—and will they spawn greater uncertainties? How will the advent of increasingly better predictions and the reduction of risk affect democratic politics? Will risk generally fade away?

We explore those questions in the coming chapters. Although we strongly believe in the long-term benefits of the Age of Prediction, we also understand that more accurate prediction, like any great advance, will create tensions and stresses, unexpected consequences, and unanticipated realities. The changing relationship of prediction and risk has already triggered tensions between security and privacy, democracy and control, rationality and irrationality. Those who master these predictive tools will have a tremendous advantage over those who don't, and manifestations of inequality, already a problem, will afflict individuals, corporations, nations, and entire regions. Improved prediction may drive economic growth and change, enriching some, impoverishing others; the effects will be powerful and intrusive.

The Dynamics of Risk

A prediction is a statement about the future. Risk is the probability that the statement is wrong. Absolute certainty, if such a thing exists, is 100 percent; absolute uncertainty is 0 percent. The probabilities of one future over another, whether it's a car weaving through city traffic, a medical diagnosis based on a genomic analysis, or the chances of one political candidate defeating another, fall somewhere in between. Certainty and uncertainty shift within that 0 to 100 percent range. Most of the mechanisms that drive predictive technologies involve calculations, based on data, that try to define the line between the two or to maximize certainty over uncertainty. The language of prediction is probability. What are the odds of a hurricane taking one path or another? What is the probability of getting struck by lightning on a golf course or of an asteroid hitting Earth? Risk is associated with failure. In finance, that failure is the trading strategy that goes wrong; in medicine, it's the therapy that kills; in government, it's the policy that fails.

How can we reconcile these two linked but very different concepts, prediction and risk? How can we be living in the Age of Prediction and still feel that risks—uncertainties—are piling up around us? Nearly all these predictive technologies involve people, and that makes the problem far more complex. Humans are not machines or only physical systems. They change their minds. They calculate based on incomplete information. They act, and then react, and then react to the reaction. They live in complex social groups. They "follow their gut." They aren't, despite the hopes of generations of economists, rigidly rational and self-interested, but they're not essentially irrational, either. The world around them never stops changing. Human consciousness is plastic, flexible, and diverse. *Human behavior, in short, is difficult to predict.*

The Age of Prediction is not an age of perfect certainty. Risk will not evaporate; it may move, change shape, evolve like a virus. For all the deep thinking about risk and management of risk, the relation between prediction and risk remains something of a mystery. Like so many aspects of our world, the relationship between these two forces is ceaselessly changing, and, in our experience, one of the key factors driving this mutability is the ability to accurately predict the effects of prediction itself.

What do we mean by that? People tend to behave differently when they are no longer fearful of the consequences. This can be either good or bad or both. It's subjective and often paradoxical. Removing a risk may be liberating because it eliminates a source of anxiety and even in some cases harm or near-certain death. We no longer worry about smallpox, tuberculosis, and polio, for instance, because vaccines have minimized their threat. We trust airplanes, anesthesiology, most prescription drugs, and smartphones despite most users lacking an understanding of how they work. It's hard to find the downsides of such advances beyond some occasional side effects.

But even as some risks can be predicted, reduced, and perhaps eliminated, others continue to emerge, like a game of whack-a-mole. From a risk perspective, the chances of dying from heart disease, cancer, and neurological disease have risen, while the threat of infectious diseases, which traditionally preyed on the very young, has receded. Overall, longer life expectancy is a good thing, a wonderful trade-off. Yet, even that advance has its complexities: the high cost of caring for older people, the economic burden of an aging society, the moral dilemmas of old age that currently affect the most advanced economies.

Many people hope that future innovations will solve these issues, but it is worth noting that optimism creates conditions that nurture and encourage risk, as the economist Hyman Minsky noted in the context of financial markets. Reducing risk can shrink incentives to behave in risk-minimizing ways—to watch your health, to care for your home and car, to behave, in that old-fashioned word, prudently. Reducing risk tends to increase the sense that you can safely take on more risk. In the markets, risk is correlated with reward.

Traders move from securities that are declining in risk and reward to those increasing in risk and reward, but traders are not perfectly rational, which can lead to poor decision-making. There is even a backlash to some predictive technologies that require intimate or private data about individuals. Some of the rise of COVID-19 antivaccination sentiment was driven by a resistance to being told what to do, a protest in defense of personal autonomy or freedom, despite very apparent risks to the antivaxxers and others. Prediction can seem to be coercive or oppressive, narrowing current behavior and future options.

Thus, predictive technologies can and do spawn unintended consequences. In fact, there may always be limits to prediction in human affairs; 100 percent certainty may be forever out of reach. We may grow confident that some targeted and quantifiable predictions are overwhelmingly likely to occur, but we may not be as certain about what the secondary and tertiary effects will be and whether they can suddenly mushroom into a threat even greater than the initial problem.

The fact that prediction and risk are inextricably linked explains why the optimism that often characterizes the Age of Prediction is also shadowed by anxiety, fear, and active opposition to what accurately predicting the future will entail. At its most fundamental level, the enhanced ability to predict creates change, which can be destabilizing and anxiety provoking. What will the change mean? How will we adjust to it? Will it release unforeseen risks, like the demons in Pandora's box? In the great debates in seventeenth-century France over whether to engage in smallpox inoculations for children—using not a vaccine but the virus itself—a common objection was that the inoculations were against God's will, *even if they worked.*

As we enter the Age of Prediction, the deepest fear among some people is that we will develop techniques to approach absolute predictability or certainty and thus render aspects of life completely determined, as if people were machines and humanity were assuming the omniscient role many once ascribed to God. A large literature speaks to this fear. Such a possibility would have bizarre effects. It would erase reward from the markets, destroying their function, and it would rob individuals of free will or at least of the perception that they can act freely. However, such an eventuality remains far off, theoretical, and perhaps unrealistic. Some quantity of risk will undoubtedly remain. As a result, we may be fated always to wonder if the very potency of our predictive tools means we are missing some new risk out there, quietly gestating in the dark.

No Turning Back

Returning to the COVID-19 pandemic, we can now ask with more detail: How *did* that pandemic occur, and could we have predicted it? Quite possibly. We now have global systems that are sequencing DNA and RNA from human patients, wastewater, and animals to find and track new viruses, such as the MetaSUB Consortium, the CDC's National Wastewater Surveillance System, the Cornell Center for Pandemic Prevention and Response, and GeoSeeq, an infectious disease tracking platform launched by Biotia (a health-tech company cofounded by Chris). Also, we now know the conditions that are driving the emergence of zoonotic pathogens, including rapid, human-driven ecological change (such as climate change and urbanization) and destruction exacerbated by the impact of increasing populations of humans living longer and traveling more. A pandemic is not just the bats' fault. Pathogens are being driven from animal and plant reservoirs where they have coexisted symbiotically (and relatively benignly) for long periods of time; once they are disturbed, they can naturally take up residence in the humans who are crowding in on them but who lack immunity against them. The phenomenon is thus a result—a paradoxical result from humankind's point of view—of our increasing success as a species on this planet leading to our increased risk. Once again, the power of human "success," much of it attributable to prediction, minimizes older risks only to see newer ones rise up, demanding ever more novel and innovative predictive tools.

3 The Quantasaurus

We represent an odd pairing: a geneticist who maps dynamics across human and microbial genomes and a quant who trades in financial markets by analyzing all kinds of data. In March 2016, the two of us were eating lunch and talking, mostly about data, but especially about how to get more of it. We were, and are, both fixated on getting as much data as possible. We often compare notes on algorithms and AI, examining the range and types of data that are pouring out of our various experiments and operations. Both of us are awash in data, and we still want more. We know that data propagate more data, which is why Igor so often talks about the possibilities of exponential increases in data. The sheer scale of the data has led to a syllogism that Chris often uses: if data give us more power to predict and more data are produced every day, then every day we wake up is the best day to be a scientist.

It is an exciting time.

Our fields, although very different, are converging around computer-driven, quantitatively based techniques for sorting through massive amounts of data in search of patterns, regularities, and signals of any kind. The jargon of the two fields is different, but the underlying tasks are much alike: both of us are in the prediction game. We are focused on the raw material of prediction, data, increasingly characterized as Big Data. We aren't alone, of course—Big Data have become a big deal. We joke about our fixation, as if it were a pet, and it has a name:

The Quantasaurus.

Quanta suggests *quantum*, as in *quantitative*, *quantum physics*, *quantum mechanics*, *quantum computing*. The word *quantum* simply means "a quantity." Igor makes a living in quantitative investing. Besides creating early video games as a teenager and working at Bell Laboratories for a while, he helped pioneer trading-based algorithms. Early in his finance career, he developed a single algorithm that sought to predict the fluctuations of stocks around market closing times; he built a business around that algorithm and, later, lots of other models (called "alphas"). One of the fields that Chris works in is computational biology, which involves using algorithms and models to untangle the increasingly complex molecular world that genomics has revealed. Both of us obsess over quanta.

Sauros is Greek for "lizard" and combined with *deinos*, or "fearful," produces "fearful lizard or dinosaur." We like the paradox of a quantum dinosaur. There once was an actual dinosaur called the Qantassaurus, a two-legged, plant-eating ornithischian that once roamed what is now Australia. The ornithischian group consisted of lizards with birdlike hips; its members may have been the forebears of birds—some grew what appear to be early versions of feathers.

Our Quantasaurus is abstract and metaphoric; it has more in common with the faux monsters Igor created in his 1980s video games than with anything that wandered around Australia chewing plants 115 million years ago. It is a product of the digital age, an amalgam of advanced AI technologies. Unlike the extinct dinosaurs slowly turning into oil, our creature is growing quickly and turning into myriad other forms and models. Exponentially quickly.

The Quantasaurus feeds off data. Much of these data race across the internet, where various electronic devices vacuum them up, package them into data sets, and pass these data sets along to customers and users, who apply them to tasks that spawn even more data. Other data may never appear on the internet but can be directly accessed through sensors (cameras, thermostats, satellites, smartphones—detectors across many spectra) or other proprietary-data-collection methods, such as surveys or polls. In the 1950s, Walmart founder Sam Walton used helicopters to count cars in parking lots, trying to judge real-estate asset values. Today, satellites can do the same but also now track weather; pollution; agriculture; deforestation; the speed, location, and direction of your car; and lots of activities people would like to keep secret. As more and more devices are linked to one another—smartphones, the internet of things, self-driving cars, weather satellites, internet searches, online videos, social media posts, GPS tracking information—the data grow larger and more diverse. Anything that can be sensed can be counted, bundled into a data set, and analyzed in multiple ways.

Whether and how data prove to be compelling rather than a random collection of observations are *the* big questions of the Age of Prediction. Vast quantities of data are not an unalloyed benefit. Data have to be cleaned and standardized to be useful. Too much data can be as much of a problem as too little, although as analytical and computational tools grow more sophisticated, data that once seemed merely "noise"—random, useless stuff—may reveal new patterns and insights. But, in general, as more data are unleashed, the time it takes for new, useful, and meaningful data to be commoditized shortens. In markets, this commoditization produces what's known as a *crowded trade*: too many users of a certain kind of data, like too many traders pursuing a certain kind of trade, drive down the value. If everyone has it, it's less valuable. In finance, you keep your best, most unique data to yourself, and get as much of it as you can. This is also true in medicine; some companies like Myriad Genetics keep all their data on cancer risk for a gene called BRCA1 from the rest of the world, so that only they can make the best predictions. In contrast to finance, siloing medical data is a detriment. As more and more people share their genomic and medical data, the data's utility continually increases for all future patients.

Two Universes of Data

How much data are out there? In 2012, IBM announced that the world produced some 2.5 quintillion bytes of data (that's 2.5 followed by 18 zeros) every day and that 90 percent of the data that existed at that time had been created in the previous two years. Since then, the world has now produced enough digital data that they now span about 100 zettabytes (100 followed by 21 zeros)—and the rate of data generation is only increasing. The advent of 5G service for smartphones is opening up vast new horizons of products and services, with ever-deeper interconnections, massive amounts of fresh data, and new forms of heretofore unimagined data in barely imaginable quantities.

The growth curve of data soars like a rocket. The mathematics of exponential growth trace a steepening curve on a graph, with big numbers being multiplied annually—not like GDP of, say, 3 percent a year, but increases of 40 to 80 percent every year (figure 3.1). That growth adds up fast.

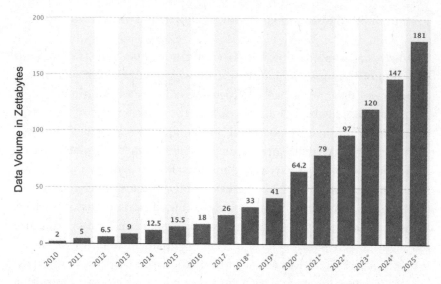

Figure 3.1

Total data created, captured, copied, and consumed worldwide. Estimates from Arne Holst show that world data production exceeded 97 Zettabytes (Zb) in 2022 and is on track to reach 181 Zb by 2025. Asterisks indicate estimates. *Source*: Arne Holst.

A famous example of exponential growth is the increasing density of semiconductor chips since the invention of the integrated circuit in 1961. Semiconductors have literally transformed our world, and they have done it at the pace of Moore's law, first stated in 1965. This law observed that the number of transistors in an integrated circuit doubles about every two years, while the cost of the chip is halved. Gordon Moore, the Fairchild Semiconductor and Intel engineer who made that prediction, updated it in the 1970s to a doubling every 18 months. Intel has said the growth of chip capacity has been slowing of late, a reality Moore agreed was inevitable, but the tech industry may yet be able to extend that capacity with quantum computing in the wings. Moore, who turned 93 in 2022, offered up another law, Moore's second, that has also proved prescient: every doubling of chip capacity doubles the costs necessary to fabricate it. Those two laws essentially define the opportunities and challenges of the semiconductor industry (figure 3.2).

The growth in data creates its own unsettling new reality. At first, it seems minor, even trivial. Data are, after all, abstract, though their effects may not be. For long periods, data grew like a gently rising pile of sand, partly attributable to an initial, very small base, partly because critical mass hadn't been achieved and explosive, self-perpetuating acceleration hadn't yet occurred. The early years of integrated circuits—which saw one transistor etched on a silicon slab become 10, then 20, then 40—may have appeared a technological feat, but it was not apparent that these improvements were changing the world. Once 10,000 became 20,000 and then 40,000, the sense of acceleration became tangible. Now that we can talk millions or billions, with components as small as a few nanometers (a hydrogen atom is 0.1 nanometer), we're seeing powerful, mutually reinforcing, accelerated growth.

We're in a new world of exponential growth of data that is, ironically, extremely difficult to predict.

Igor operates a business, quantitative investing, that is experiencing exponential growth. He has lived with the symbiotic relationship between the growth of data and the growth of predictive algorithms (his alphas) fueled by those data. At the operating level, he has amassed large numbers of alphas—from a handful in the 1990s to millions today—that are used to construct tradable portfolios. Because alphas use data that evolve with time, they continually generate more data. Everything is automated, and the performance of individual alphas is continuously monitored.

Exponential growth produces a distinctive psychological dynamic. In enterprises shaped by exponential growth, stakes rise, and competition intensifies. Exponential thinking means living with risk in new ways. The world that exists today will be dramatically different tomorrow. There is a ruthless meritocracy to this dynamic, which in turn requires a deeply empirical, flexible mindset: keep your eye on the goal, but be willing to cut your losses. It means knowing that what is shocking today may be routine tomorrow and obsolete the day after. It requires a belief that limits are meant to be broken. In an exponential world, the terrain ahead is always unknown. Turbulence is inevitable. Old risks recede, while new, unknown uncertainties may emerge.

Power Laws

Genomics is the new kid on the exponential-growth block, having ridden an explosive growth in data since the mid-2000s and the development of massively parallel next-generation DNA sequencing (NGS) methods at Solexa, which Illumina acquired in 2006. That development is one of the

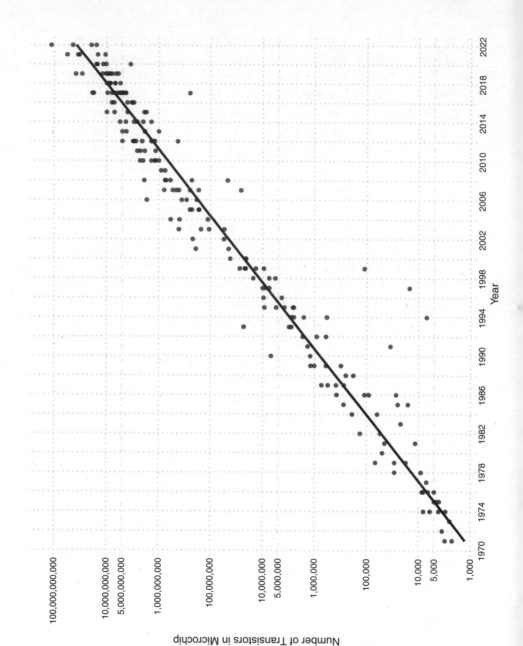

Number of Transistors in Microchip

Year

most important, if relatively unknown, keys to the Age of Prediction. The growth rate described by Moore's law has held for more than five decades, but the pace of NGS has advanced even more quickly. NGS has produced precipitous decreases in the cost of DNA sequencing and analysis that have averaged 50 percent reduction *every five months*. The falling cost of sequencing, the increase in speed, and the growing number of NGS instruments mean that each day sees a new record for the total amount of sequence data. It now costs less than $200 to sequence a 3 billion base-pair genome in under eight hours. That represents a millionfold drop in cost from the first sequencing projects, which took about a decade to complete, and a 10,000-fold increase in speed. And newer technologies from companies such as Illumina, Ultima, Element, BGI, and Oxford Nanopore are driving the cost to less than $100 per genome. Every day is the best day to be a geneticist.

The sheer volume of genomic data threatens to exceed even astronomical data scales. The genomic sciences generate vast amounts of data by exploring everything from the far reaches of the universe to the quantum world to the intricate machinery of the cell. Astronomers held previous records for "large-scale data analysis," producing exabytes of raw data. (One exabyte equals 10 to the 18th power number of bytes, well deserving of the term *astronomical*.) Genomics, however, can now create a yottabyte (1 trillion terabytes or 1 million exabytes) of data, from things as small as the RNA molecules in a single human cell to a genome for all 7.9 billion individuals on Earth. Projects such as the UK Biobank and other national initiatives have now sequenced millions of human genomes and linked them to long-term medical records, and this also accelerates the data growth.

Human cells also possess genomical-scale complexity, which massively multiplies the available data. In a cell, RNA copies of specific DNA sequences can be modified and spliced, transported to the ribosomes, then

Figure 3.2
Moore's law. The line corresponds to exponential growth with the transistor count doubling approximately every two years. The *y* axis is in log-space. Significant chip types are labeled by each data point. *Source*: Hannah Ritchie and Max Roser, "Technological Change," Our World in Data, 2013, https://ourworldindata.org/technological-change (CC-BY).

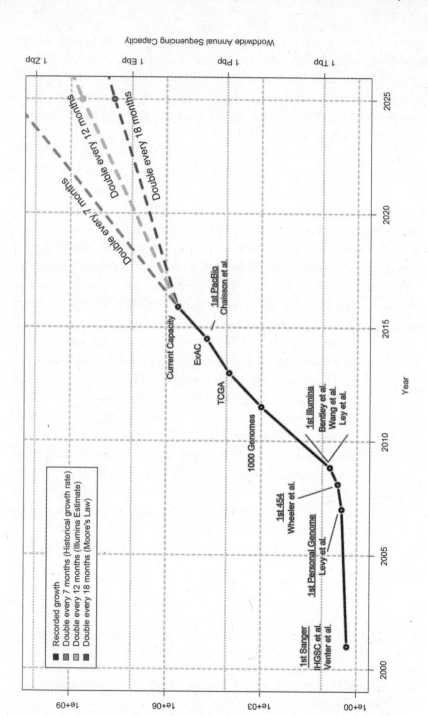

used to "program" the assemblage of proteins. The human genome has roughly 21,000 genes that code for proteins, but that doesn't begin to cover the underlying complexity. Genes can be spliced into a wide array of forms, much like mixing and matching a large set of standardized Scrabble letters to make different words. Each subunit is called an "exon," and although some genes contain just one or two exons, others have many more: the largest gene in the genome—called TTN, or Titin—has 363 exons. Because exons can be shuffled and rearranged in potentially any linear order, the number of possible splice variants—different DNA sequences—for a human being is staggering, totaling 9.4×10^{108} combinations. As a comparison, astronomers estimate that there are about 8.2×10^{81} atoms in the entire universe. Thus, strikingly, the inherent combinatorial complexity of DNA sequences within a single human cell is larger than the number of atoms estimated in the entire universe.

Because all living organisms use DNA and RNA, far more than just human DNA and RNA is being sequenced, analyzed, and modeled. Organisms used in biological research include mice and fruit flies as well as bacterial, viral, and fungal strains in clinical labs. The science of all these genomic data is known as *metagenomics*. Each day builds up more data and accretes a larger catalog of genetic knowledge, layering information from different species, biological states, and locations.

But just because you have access to genomic-scale data doesn't mean they're organized or processed—or that they're immediately useful. Many of the scientists in Chris's lab have observed how messy biological data

---

Figure 3.3
Growth of DNA sequencing and data. The plot shows the growth of DNA sequencing both in the total number of human genomes sequenced (*left axis*) as well as in the worldwide annual sequencing capacity (*right axis*: Tera-base pairs, Tbp; Peta-base pairs, Pbp; Exa-base pairs, Ebp; Zetta-base pairs, Zbp). The values through 2015 are based on the historical publication record, with selected milestones in sequencing (from the first Sanger genome through to the first PacBio human genome published) as well as three exemplar projects using large-scale sequencing: the 1000 Genomes Project, aggregating hundreds of human genomes by 2012; the Cancer Genome Atlas (TCGA), aggregating more than several thousand tumor/normal genome pairs; and the Exome Aggregation Consortium (ExAC).

are compared with financial data, much of which has been gathered and standardized for decades or longer. Ideally, raw data become organized information, which is then processed into predictive knowledge, driving an appropriate action. However, data in genomics or anywhere else rarely attain that state.

That said, markets and biology share deeper affinities that large amounts of data can sometimes reveal. Both consist of many interacting individual agents that often appear to be acting randomly but display evidence of self-organizing behavior. You can see this tendency in flocks of birds, in mud-slides, in turbulent flows of liquids or neutrons, and in the weather itself. Organization emerges from simple agent interactions and from bottom-up rather than top-down control. These processes don't require a master con-troller to operate. Evolution itself is one of the most powerful decentralized self-organizing processes, "directed" by what the evolutionary biologist Richard Dawkins famously called the "blind watchmaker," without discern-ible goals beyond survival, but determining many other natural systems and shaped by them in turn. Feedback is among the most powerful aspects of complex self-organizing processes and a driving force in both exponen-tial growth and systemic breakdowns.

These self-organizing systems often display processes that, despite the appearance of disorder and even chaos, grow at a rate based on a so-called power law. A power law is a fixed, often exponential relationship between two quantities in which a change in one quantity results in a proportional relative change in another. Perhaps the simplest power law is the fact that if you double the length of the sides of a square, its area quadruples; if you triple it, the area grows by nine times, or three squared. Data today are growing by a power-law exponent over time. Moore's law clearly defines a relation between time and semiconductor capacity. Mathematically, if you graph a power law, you produce a straight line with a slope (positive or neg-ative, rising or falling) that fixes the relationship between two quantities.

Nature is rife with these regularities. Metabolism operates more efficiently as animals grow in body mass (by a 0.75 exponential increase), and aspects of urban infrastructure—length of roads, electrical cables, water pipes, the number of gas stations—scale in the same way in cities of very different sizes. At the start of the twentieth century, the Italian economist Vilfredo Pareto articulated the now-famous 80–20 relationship, which defines the allocation of wealth in a society: 20 percent of the population controls 80 percent of the wealth (or, in a more popular formulation: 20 percent of the people do 80 percent of the work). Since Pareto articulated this general distribution, it has been found to apply to a number of economic, sociological, and natural phenomena: the distribution of cities by size in a country (fewer large cities and many smaller); the values of oil reserves (a few large fields and many small); hard-disk-drive error rates (a few large errors and many small); companies by size (a few very large companies and many small). Like Moore's law or the rate of genomic sequencing, the exact Pareto relationship may drift—say, in income allocation—but the notion that a few big entities (cities, meteorites, the wealthy, genomes, the hardworking) exist in a fixed relationship with a much greater number of smaller or lesser entities appears to have the straight-line slope characteristic of a power law. Power laws, in short, are predictive.

The Foundation of Finance

Finance has long been a matter of prediction and has extensively used power laws and other tools. Economists describe markets as pricing mechanisms, but at the heart of the pricing mechanism is a trade, a bet, a gamble, a belief on a future price. Markets are probabilistic and prospective. Markets are also leading indicators, with the shortest-term traders and the longest-term investors trying to gauge the future of everything from the growth of the global economy to the path individual stocks and bonds will take.

Early in the twentieth century, the attempt to predict future prices took on a more scientific air with the introduction of forecasting and statistics. Forecasting pioneers such as Roger Babson, John Moody, Yale's Irving Fisher, and Columbia's Wesley Clair Mitchell developed statistical systems to try to predict the future course of stocks. Though such systems were popular throughout the 1920s—many were sold on a subscription basis to investors—nearly all of them failed to predict the crash of 1929 and the Great Depression. Fisher, in particular, famously declared that "stock prices have reached what looks like a permanently high plateau" just before the crash. He, at least, had tenure to fall back on.

Although the Great Depression battered the reputations of market forecasters, it also drove academic research to explore market processes and dynamics; that research began to produce results after World War II. Much of the work on markets and investing occurred in and around the University of Chicago from the 1950s to the 1970s and involved an attempt to build a stronger, more quantitative foundation for investing. That movement, now known as Modern Portfolio Theory (MPT), included assumptions, techniques, and hypotheses that continue to define mainstream investment practice: efficient and rational markets, the random walk of prices, portfolio diversification, volatility, and arbitrage. MPT began in an era when calculations were still done with slide rules and mechanical calculators, but advances in computer technology soon came to play an increasingly major role. And, as we've seen, the ability to crunch more and more numbers drove an explosion of data, producing powerful feedback loops and igniting exponential growth in data.

That didn't mean that markets had suddenly grown predictable. Markets are complex and elusive. As Emanuel Derman, a physicist turned Goldman Sachs quant, wrote in 2002: "There is no fundamental theory in finance. There are no laws. Models in finance are used to turn opinions into prices, and prices into implied opinions." Finance is not physics, Derman emphasized, but more like social science. "Models in social science are . . . toy-like descriptions of idealized worlds that can only approximate the hurly-burly, chaotic and unpredictable world of finance and people and markets."

Nevertheless, the new thinking about investment did provide a way for investors to attempt to balance two attributes that have long been known to shape any investment: risk and reward—the greater the risk, the greater the expected reward, and vice versa. Market practitioners have long understood "risk and reward" as a general concept that is all but impossible to quantify. There's no certainty, for instance, that by adding a fixed quantity of risk you will get a fixed quantity of reward; indeed, anyone who says so would be considered hopelessly naive. Rather, increasing risk suggests that your reward may be large, but it may also be small or nonexistent or a brutal loss.

That's because risk embodies a quality of future price movements known as *uncertainty*. Like any attempt to predict the future, investing is laden with uncertainty, a factor defined by the University of Chicago economist Frank Knight in 1921 as a risk that can't be rendered as a probability and therefore can't be predicted. Much of what MPT did was to begin to define—quantify—risks that were not in the realm of uncertainty. Investors had to take some risk to generate some reward, and with MPT they could in theory identify, quantify, and minimize risks to tilt the probabilities of reward in their direction. Yet, MPT hardly provided the secret of accurate prediction. Even though US stock markets were viewed as increasingly efficient—they adjusted to changes in new information quickly and made beating the market very difficult—they continued to be vulnerable to sudden and unpredictable drawdowns, bubbles, crashes, and devastating losses.

MPT Going Digital and the Birth of AI

Beginning in the 1930s, developments were unfolding outside the world of investing and markets that would change how those risks and rewards were quantified. Like so many aspects of digital computing, the story begins with Alan Turing, a British mathematician, cryptologist, and logician. In 1936, just as the possibility of modern computing was emerging, Turing laid out in a paper titled "On Computable Numbers" the basic functions that such machines would require. Turing theorized about a machine that "could read, write, remember, and erase marks on an unbounded supply of tape," as George Dyson writes in *Turing's Cathedral*. Such a machine became known as the "universal Turing machine," a device that, given sufficient time, tape, and description, could simulate the behavior of any other computing machine with the same attributes. Turing laid out these requirements for digital computing without regard for the underlying technology. "Being digital," he wrote, "should be of greater interest than . . . being electronic."

As computers and data grew exponentially, computer and data science developed new tools and techniques to exploit that growth, and from that development came the broad field popularly known as *artificial intelligence*. Humanity had speculated about the possibilities of "machines" that could think like people since the ancient Greeks, from Pygmalion's statue in Ovid's *Metamorphoses* to Leonardo da Vinci's humanoid robot to Thomas Hobbes's Leviathan, an "artificial human" that embodies his ideal state. Dyson once described Hobbes as "the patriarch of artificial intelligence."

The first serious work on AI began in the 1950s as computer technology rapidly advanced. The phrase *artificial intelligence* was coined by the Dartmouth College mathematics professor John McCarthy in a proposal to the Rockefeller Foundation for a summer meeting in 1955 that would bring together the leading lights of the nascent field. McCarthy had invented the name to distinguish AI from what he considered narrower pursuits, such as automata theory and cybernetics. Over the decades, however, AI has been plagued by an often-confusing tangle of definitions—the symptom of an open-ended and fast-moving field. As one commentator noted, the problem was less in the term *artificial* (though some early AI researchers associated it with "phony") than in the term *intelligence*, which suggested machines that could think like humans. Particularly in those early days, computers were often described as "giant electronic brains" or "thinking machines," though they were actually just fast (for the time) calculators and filers. But what *is* intelligence? Is it the ability to recognize and distinguish human faces or handwriting? Is it skill at playing complex strategy games or solving difficult questions? Or is it the ability to learn?

Turing played much the same kind of role in AI that he had in digital computing. In 1950, he wrote a paper, "Computing Machinery and Intelligence," that attempted to define an intelligent machine. The issue, he contended, was not whether the machine was infallible and correctly answered every question submitted to it. The Turing test of machine intelligence was whether a human interacting with a computer through an exchange of texts could not reliably tell whether it was human or machine. Turing, who believed machines could learn, made provocative proposals for what he called a "learning machine" and discussed the possibilities of "evolutionary computation." He had a keen sense that evolution could expand beyond biology.

In contrast to the early, limited machines, AI in the Age of Prediction is a vast, diverse, and swiftly moving field with a rich interplay between computer science and many other fields, from materials science to medicine. Although most predictions concerning AI continue to improve with more data, the history of AI is also a graveyard of failed predictions. In a 1957 talk, Herbert Simon, a Carnegie Mellon economist and AI pioneer, predicted that in ten years machines would be champion chess players, compose music, prove a mathematical theorem, and embody a psychological theory as a program. He said, "Machines will be capable, within 20 years, of doing any work a man can do." Another AI giant, the MIT computer scientist Marvin Minsky, predicted in a 1968 press release for the movie *2001: A Space Odyssey* that "within a generation . . . the problem of creating 'artificial intelligence' will substantially be solved." Neither prediction came true on schedule. AI experienced booms and busts, including a stretch in the late 1980s and early 1990s dubbed the "AI winter," when many practitioners lost heart and funding was scarce. Lots of excitement had been generated in the 1970s by programs known as expert systems that tried to capture the expertise of, say, physicians, accountants, and lawyers in computer programs. But expert systems were inflexible—mechanical.

Expert systems never regained their preeminence, but other techniques of machine learning, such as neural networks, came to the fore. For example, the growth in computing power enabled the development of larger neural networks with more layers and connections for identifying subtle patterns in data. Data became increasingly important. Expert systems were programmed to provide logic to the machines, which processed the instructions fed into them. The *instructions* were essential, not the data. Computer scientists and programmers began to design and build computer architectures that didn't just respond to instructions but "learned" from data, extracting patterns, logic, and signals from rich, deep, and increasingly diverse data sets.

Today, AI takes many forms, including machine learning, deep learning, natural-language processing (NLP), expert systems, and fuzzy logic. The first, machine learning, is a way of programming learning based on test and training data and includes supervised, unsupervised, and reinforcement learning. For more complex data, deep learning implements variations of neural networks that resemble aspects of the human brain, processing high-dimensional data to seek emerging patterns. NLP involves learning patterns of word interactions and vectors. Amazon employs it to understand patterns in customer reviews and to improve user experiences on its website, while academics have shown that NLP algorithms can even be used to detect crime-related tweets on Twitter. Expert systems, although less common today as independent programs, are part of almost every programming language. Fuzzy logic uses program parameters that estimate the probability of an answer being correct, as opposed to the simple "yes/no" of Boolean logic.

Fortunately, the AI winter began to thaw in the 1990s with a handful of splashy promotional triumphs. In 1996, IBM's Deep Blue, initially developed at Carnegie Mellon, defeated the chess champion Gary Kasparov, belatedly affirming Simon's prediction of 1957. In 2011, IBM used a question-answering system—a combination of an information-retrieval system such as a structured database and NLP—called "Watson" to defeat several *Jeopardy!* champions. And in 2014, DeepMind Technologies published a description of a program called AlphaGo that could play the Chinese game Go. Though Go's rules and board are simple, the game is difficult to master. Because each 19-by-19 location on the square board can be either empty, black, or white, there is a total of 3^{361} possible board positions and more than 10^{170} board configurations. In March 2016, AlphaGo, using an advanced search tree with deep neural networks, played Lee Sedol, one of the world's best professional Go players. AlphaGo won four out of five games.

AI for Finance

In 1995, Igor took a job as a portfolio manager at a New York City hedge fund firm. At the time, mathematically sophisticated, academically trained professionals were popping up on Wall Street, and "quant finance" was emerging. Though markets often appeared to be chaotic, research was turning up persistent regularities. Some were stronger and more persistent than others. Igor built his first algorithm and over the next decade evolved his own set of ideas about the most effective way to use his "alphas" to try to predict prices. Much of this work involved scaling up the number of alphas into what he began to call an "alpha factory." In 2007, he formed his own firm, WorldQuant.

Just as our immune system keeps learning about evolving microbes, Igor recognized a similar reality of market regularities: no regularity lasts forever, so there's a constant need for new alphas to identify new market signals derived from new data sets and new insights. Over time, he developed what he calls "tenets of prediction," which laid out a kind of mathematical circularity among models, data, and predictive accuracy. He found that with his strategy the accuracy of models tended to rise logarithmically with the number of algorithmic models, so 100 times more models produce, in theory, 10 times better predictions.

WorldQuant is in the business of trying to predict stock prices and market trends. Every model or formula attempts to predict returns based on a very specific piece of information or a very specific way of looking at a piece of information. For example, WorldQuant has several hundred thousand alphas built on relatively simple and publicly available price and volume information. One alpha may look at five-day returns, another at one-day returns, yet another at returns relative to similar stocks. Each different way of looking at data yields a slightly different prediction, which, in turn, interacts with other predictions. Broadly speaking, if combined predictions are uncorrelated, Igor concluded that they increase their power with the square root of the number of predictions. If they are correlated, the increase is less steep, typically following a logarithmic improvement for each new alpha. (The shorthand of logarithms is an essential tool in dealing with exponential growth.) But if you combine an exponential number of logarithmic improvements, you get a linear improvement with the logarithm of alpha quantity. That's why Igor believes it is an advantage to have millions of alphas. These alphas can reveal a signal that is much stronger than any of the component signals, which cannot be described by a simple formula.

This is a complex endeavor in practice. To predict something based on different data, you can mix the data and then try to predict, or you can make predictions based on each alpha derived from each data set. The latter is much simpler, especially at scale with thousands of data sets and millions of predictions. It's not the amount of data but the number of *quality* models built from the data that ultimately drives prediction in a strategy like WorldQuant's. But the raw number of models is proportional to the number of useful fields in any given data set. This prediction strength is proportional to the logarithm of the volume of quality data, which explains why exponential growth in data is required to keep prediction improving linearly. Quantity is necessary to win the prediction war in the markets, in the immune system, and in any predictive system.

But this begs the question, why do signals come and go? First, markets are fluid and prone to continual and often random change. That fluidity isn't just a metaphor. Quants borrow a term from physics that describes a turbulent, nonlinear flow—*stochastic*, which means the flow is somewhat random and therefore can only be analyzed statistically. Markets are clearly nonlinear, and prediction is almost never a black-and-white binary choice, but rather a probability. Second, profitable strategies attract imitators, which may flatten out that probability and the associated profits—a remorseless process in financial markets known as *arbitrage*. This is why Igor uses the metaphor of a "prediction war" and why getting data is effectively an arms race.

That arbitrage often comes from rivals. Few really good ideas go unnoticed, and if enough professional traders, ceaselessly hunting for mispricings, show up at the same place at the same time, a strong signal will get smothered and leave everyone disappointed, like when everyone goes to the same good restaurant at the same time. More subtly, an organization can find itself inadvertently moving in sync, like sheep in a flock blithely heading over a cliff. Algorithms may contain internal biases in approach that their developers never recognize. A bias can create inadvertent correlations in seemingly diverse strategies, spawning a crowded trade; if such unconscious herding is large enough, it can result in serious losses that spill throughout the market and into the larger economy. The result is that specific strategies suddenly cease to generate profits or that entire markets suddenly decline. These biases are psychologically difficult to recognize or eliminate, and they may be deepened by the overconfidence that you are beyond bias.

Idea Arbitrage

Let's continue looking at one of the big ideas that shaped MPT: arbitrage. University of Chicago economists had long argued that the ceaseless activities of self-interested professional investors established the market's essential nature as efficient and rational. As skilled consumers of financial information—in contrast to what the economist and proto-quant Fischer Black, one of the creators of the Black–Scholes options pricing model, referred to as "noise" traders, who operate irrationally or emotionally—these investors are continually searching for market irregularities: stocks that are mispriced, such as two securities with the same underlying asset but two different prices trading on different exchanges. These traders engage in arbitrage—for example, by buying, or going long on, an underpriced security and by selling, or going short on, an overpriced one they bet will fall. Arbitrage is a fundamental aspect of trading, but devotees of the Chicago school viewed it even more broadly as the engine driving the market toward accurate prices—price discovery *itself* as a form of prediction—thus rendering financial markets efficient (quick to react to changes in information) and rational (driving a securities price to its intrinsic value or true equilibrium price).

The power and centrality of arbitrage have always provoked debate. But the essence of arbitrage is really just the power of competition to produce a winner. By analogy, Darwinian evolution is a competition as well, with survivors representing not greater perfection but expedient adjustment to current conditions; markets are always about current or future conditions. Gather enough agents to freely compete, you may well generate better results than a lone trader, even if that competition isn't always efficient, rational, or accurate. Igor's alpha factory was designed to be flexible and expedient and to aggregate many different signals.

In 2018, Igor started the company WorldQuant Predictive to apply the same ideas beyond finance, and to commercialize prediction in other industries. The company used the same basic approach of iterative testing and competition—"idea arbitrage"—as WorldQuant. To predict more effectively, you need algorithms that approach the problem from many different directions. To do that, you need model builders who think creatively and data sets that offer as many different perspectives on a problem as possible. Scale is essential. Through iterative backtesting, the ideas embodied in algorithms compete with one another. Arbitrage occurs, and the best-performing ideas emerge. All this was baked into a formula: ideate (develop ideas), arbitrate (the verbal form of arbitrage), and predict. The same process works for aggregating cancer data mutations, drug combinations, and patient outcomes or even for tracking a novel infection; the models continually need to improve and be adjusted based on the latest data.

Again, Igor's strategy of amassing large numbers of alphas is just one of many processes that have emerged in the race to make more accurate predictions. The operations of any of these prediction businesses combine human talent, powerful machines, and fire hoses of data. Moving beyond finance to a more diverse set of industries, which is what WorldQuant Predictive and a number of other companies are doing, is not necessarily easy. Health care, transportation, insurance, telecommunications, and consumer products are very different from finance, with its rich historical and market-pricing data, high transaction volumes, fixed numbers of instruments, and many, if weaker, signals. Each of these industries poses different problems, operates with different dynamics, requires different data sets, and possesses different limitations. And it remains an open question as to how many alphas or signals are necessary to build a strong-enough library of predictive algorithms for each one. We learn by doing.

An algorithm, for instance, may be built around the predictive notion that a tech stock that goes up three days in a row is likely to go up on the fourth. That signal may be relatively weak, though: it may be true some of the time, but you wouldn't necessarily want to bet big on it. But if you get hundreds or thousands of such signals operating together, you might be able to build a model that is right 55 percent of the time, which may be all you need as an investor.

Of course, 55 percent will not work particularly well if you're making predictions about health outcomes or even trends in, say, grocery shopping. Take a hypothetical insight that could emerge from survey and supermarket data: "A mother with two children buys apple juice only on Wednesdays." You probably wouldn't want to bet much money on that. But you can use machine learning across many predictions and "ensemble" them—the term comes from statistics and means a probability distribution of the state of a system that is combined from several models—so that they all are merged into a production model. The output of signals, which themselves are models, produces inputs to generate even stronger models and better outputs. For example, for predicting which species (out of the possible millions) is causing an infection, ensemble methods in genomics can combine the best features of various models to give the best prediction, as was shown in a paper from Chris's lab in 2017, "Ensemble Approaches for Metagenomic Classifiers."

An important benefit of the "ideate, arbitrate, predict" process and the ensemble method is that you can integrate signals from very disparate data sets. For instance, you might not think to integrate weather data with credit card data, but you can build ensemble models that come from weather data and credit card data to predict the products people might buy when the weather is nice rather than, say, cold and rainy. In influenza and now COVID-19 models, researchers in Chris's lab and others now integrate epidemiological data with weather data, transportation data, and data from proxies for social distancing, such as mobile phones. Then predictions can be built to discern whether an area is high risk or low risk for transmission of a virus, such as SARS-CoV-2, but the same models can also be used for other infectious diseases.

Proxy Predictions

The fuel for prediction remains data. The diversity of data types, especially when they are combined with today's algorithms, effectively reveals developments that were formerly invisible, such as viral movement and evolution. The accuracy of prediction grows with a logarithm of the amount of data that one can access, meaning that more and more data are needed to predict things more accurately. But with data growing exponentially—and there's no evidence that this state of things will soon change—predictive power is growing increasingly accurate.

One key to predictive performance is to access not just publicly available data and client data, but a kind of secondary data spawned in a world that's ever more densely interconnected. This *proxy data* can often provide striking new insights from unusual sources. For example, in the early days of the COVID-19 pandemic, Kinsa, a company that sells "smart" digital thermometers, was able to upload data from millions of its Bluetooth-enabled devices to create a map of elevated temperatures across the United States that was updated in real time. Kinsa was not directly measuring the spread of the virus, but by measuring abnormally high temperatures that could be associated with the viral infection, the company was able to produce a real-time sense of the evolving spread—an important service, given the paucity of testing at the time.

In 2009, the United Nations launched Global Pulse to bring data-based prediction to policy making, which up to that time had relied on traditional surveys and statistics. Led by Robert Kirkpatrick, a pioneer of software development for humanitarian purposes, Global Pulse engages in what's known as "global data philanthropy." Companies gather data, and Global Pulse taps into those data to extract real-time predictive insights. In Rwanda, the amount of money people spend on mobile phones can predict with 89 percent accuracy how much market-dependent households are allotting for food. In effect, every mobile operator in Africa is running a food-security-monitoring network without knowing it. Mobile phone records can reveal where people have been displaced by disasters or where they're assembling in need of assistance or when warfare has started between sects, and such models can be used to understand where supplies need to go. When you relate the movement of a population hit by disease outbreaks with information on vectors and rainfall, you can feed that into epidemiological models, which can be powerfully predictive and helpful. Purchasing trends can even predict when a war has started because noncombatants begin to purchase staples and basic supplies in advance of troops, as the Global Pulse data have demonstrated.

Other examples have already been piloted and are being deployed. For example, shipping data in the Mediterranean can detect migrant rescue events and predict where refugees are likely to come ashore. Deep learning models can be trained to recognize those patterns, so authorities can be prepared to receive people before they arrive and help them or redirect them. Similarly, xenophobic and hate speech can be mapped by mining Twitter and Facebook pages using NLP. Researchers can use data from companies such as PayPal, Walmart, and Amazon.com to create a real-time map of the progress of every neighborhood in the world in, for example, the transition from incandescent to LED lighting, from dining-room sets made of rain forest mahogany to those made from bamboo, from equipment used to upgrade old heating and cooling systems to new, energy-efficient technology.

Proxy data are hardly perfect, particularly in the use of diverse and poorly understood data sets. In statistics and data science, there's a concept known as overfitting, in which researchers fit a function too closely to a limited set of data points so that they are led to believe the results will be more predictive than they turn out to be. Indeed, one subtle temptation in modeling is to pluck from a sea of data only those data points that confirm your already-existing predilections and biases. Proxy data are, by definition, limited. You're measuring not the primary reality but an inferred secondary characteristic. What if those Kinsa temperatures are not from COVID-19 but from the flu, or from the density of people who like to run outside when it's hot? All models have imperfections, and these imperfections have to be balanced and normalized as well. As the statistician George Box famously said, "All models are wrong, but some are useful."

However, all those limitations are just part of the modeling. Through arbitrage, which can test, model, and ruthlessly compare—separating true from false and fantasy from realistic prognostication—it is possible to account for such risks and errors. Indeed, this leads us to the question of risk: the trade-offs, paradoxes, and unanticipated feedback loops that may emerge if prediction actually works.

4   The Trouble with Risk

It was a crisp December morning, and Igor was having his DNA sequenced. Most people would buy a kit from, say, 23andMe or have a test ordered by their physician. Igor asked Chris.

DNA analysis today is relatively straightforward, comes with many predictions, and has become common; millions of people have had their DNA profiled. This is no small feat because the genome is big—if a typical human genome were printed and bound into a book, it would require 130 volumes of text consisting of only four letters (A, C, G, T), with no spaces, syntax, or obvious sentences. A reasonably fast reader would need about 95 years to read what would appear to be a random script about these four nucleotides. Sequencing technology is much faster; a whole human genome can now be sequenced, analyzed, and clinically applied in just a few hours.

Genes are fundamental elements of heredity, some quite small (less than 1,000 letters, or bases, long), others large (more than 100,000 bases). Many encode for specific proteins, but many are noncoding and have their own functions or operate in concert with protein-coding genes. Some are mysterious, yet the genome is not uniformly random. Rather, within seemingly random sequences lurk structural elements of biological prediction; there's a code that we can read.

In the popular culture, genes translate into traits. You're smart because you have "smart" genes; you sing opera because you inherited a set of "singing" or "musical" genes. But, in fact, the translation of genes into a phenotype (trait) is far more complex than that. All genes are activated, modified, and attenuated in response to changes in the environment; they are controlled by *epigenetic* mechanisms: these mechanisms operate "on top" (epi) of the genetics—they alter the operation of DNA without changing the code. For instance, the four letters of the genetic code are sprinkled with chemical alterations in which a methyl group attaches to specific bases and, in some instances, helps repress a gene's activity. This process enables the single genetic code from the first single cell in every human embryo to modify its functions and genes' activities to fit the needs of the eventual trillions of cells in the human body.

What genes do is almost magic, and it comes with a set of its own predictions, too. Specifically, every human genome offers a glimpse into a possible future of a range of illnesses, disabilities, and cellular disorders as well as abilities and putative "genetic superpowers" that may protect us from some diseases, such as HIV, or reduce the risk of others, such as heart disease or cancer. We are dealt shuffled cards from the DNA of our parents, each of whom contributes half the genomic deck, which we then shuffle again into the next generation.

In a genomic analysis, we often isolate the DNA and use what's called a "shotgun-sequencing" approach, fragmenting the chromosomes into pieces (like shooting with a shotgun) that are around 300 to 500 bases long. We then align, or map, fragments of DNA to the known human genome and look for differences. Small differences of only one letter—a single base— are called "single-nucleotide polymorphisms" (SNPs) or "single-nucleotide variants" (SNVs). But there are also larger variants, such as insertions or deletions (indels), entire chromosome-size changes, extra sections of DNA, and even extra chromosomes, as in Down syndrome, which is marked by an extra copy of chromosome 21. All of the variants of a genome are then compiled, mapped, and summarized, which is what Chris did for Igor.

Altogether, Igor had 5,031,353 genetic variants, which is not unusual. Every human is about 98.7 percent genetically the same as other humans; family members are even closer. The big question is which of the 1.3 percent of differences, the variants, matter the most for health and disease. This is a fundamental question in what's become known as precision medicine and is relevant for every person: Can *my* particular mutations predict which health conditions I might develop in my lifetime and which drugs or treatments I can and should receive?

Genomes are evidence of ancestry. The color of our eyes, skin, and hair as well as subtle facial and body features are somewhat defined by our ancestry, as are the ability to process lactose in milk, variations in blood types, and vulnerabilities to various diseases. The genome of each person must be understood in the context of human ancestry. Igor's ancestry showed over-all eastern European origins, with a clustering of Bulgarian Jewish origins. Indeed, this matched his Jewish heritage.

One mutation in Igor's genome matched a reference SNP called rs12913832 that is a link to eye color in a gene known as OCA2, or oculo-cutaneous albinism II. Igor had inherited a G allele (a gene variation) from both parents, meaning he was homozygous for this allele—that is, both copies of the gene were the same; if he had two different versions, he would have been heterozygous for that trait. This homozygous variant and his European ancestry gave Igor a more than 99 percent chance of having blue eyes, which he does.

Igor was homozygous for another SNP, rs762551, which meant he was a fast metabolizer of caffeine. Other variants suggested he had a better than average ability to taste bitter foods and a reduced risk of migraines, was likely to be a sprinter (versus a long-distance runner), and likely had higher resistance to norovirus infections.

Some variants illuminated Igor's choice of careers, but making this link took us into more speculative territories. One of his variants linked to increased susceptibility to "novelty- or risk-seeking." These mutations regulate dopamine D4 receptor levels. Dopamine is a neurotransmitter that, when released, makes us feel happy, satisfied, rewarded, or on a high. In some people, there appear to be fewer receptors that regulate dopamine production, leading to a predilection for risk; these people need to take on more risk to experience the euphoria that dopamine produces. The willingness—indeed, the eagerness—to engage in risk extends to many areas, from sports to personal behavior to businesses that involve risk-taking, such as gambling, trading, auto racing, and sky diving. Scientists have speculated that addictive behavior results from a surfeit of dopamine-inhibiting receptors, which produces a craving for dopamine. Other studies have shown a possible link between risk engagement and the density of dopamine-transporter proteins, regulated by another dopamine gene.

Igor also had a variant associated with faster learning in so-called Go versus No-Go situations that is also associated with the dopamine receptor gene. Subjects with this mutation tend to exhibit better episodic memory, helpful for making rapid decisions. Clearly, risk-taking and its rewards and punishments are important elements of learning, competition, and evo-lution. Some studies assert that human evolution has positively selected overall risk-taking behaviors as well as behaviors once associated with mass migrations, competition, and, potentially, cooperation. These mutational elements are fluid over time among a population, just as they are among individuals. They also involve an interaction between nature (the reality of individual dopamine systems) and nurture (the complexity of human interactions with the social and physical environment).

The fact that Igor is comfortable with risk shouldn't come as a big surprise, given his job. Yet the overall likelihood of someone having a predilection for risk-taking is built on a predictive model that, itself, has some risk of being inaccurate. Indeed, a genomic analysis captures the two basic concepts that shape the Age of Prediction: prediction and risk. Both are probabilistic; what they reveal provides possibilities, not certainties. Straightforward predictions are rare. And what is risk? To answer that, we have to start with the nature of luck.

Luck and the Modern Era

What is luck?

Luck is success brought on by chance rather than through one's own actions or skills. Metaphorically, luck is a winning result in a game of chance, such as a game of dice. Tumbling dice display a random quality, and the gambler has no choice but to call on some superstitious power or totem—a rabbit's ear, a lucky coin, a kiss from a stranger, a four-leaf clover. Luck exists in some metaphysical netherworld where stars and the moon can shape events. Luck can send the gambler on a hot streak or to disaster; luck can enrich or impoverish. It is, after all, impossible to predict what numbers will come up when the dice hit the green felt and roll.

In contrast, games of skill involve mental or physical prowess. In games of chance, all the skill in the world can't guarantee success. In fact, some skills, such as card counting, are banned by casinos. But gamblers have developed skills that, although technically not predictive, can tilt the odds in their favor. Indeed, games of chance, with their dependence on luck, provided the seedbed of probability and statistics, two symbiotic sciences that revolutionized how we perceive and handle random phenomena, from dice to market prices to observational errors to the shuffling of parental DNA at conception. These twinned scientific disciplines, which emerged in their modern forms in the sixteenth and seventeenth centuries, transformed the landscape of prediction—at first a little, then a lot—by exposing regularities and order beneath seemingly random change. At the same time, they led to the ability to define and quantify the risk of every prediction.

As the eighteenth-century Scottish philosopher and historian David Hume wrote, we can never be absolutely certain that the sun, having set the night before, will return in the morning. We believe that will be the case, and each day confirms the case a tiny bit more, but there always exists that narrowing sliver of uncertainty, or risk, that it may not return. As the stockbrokers say, "Past performance is no guarantee of future success." This is known as Hume's induction problem, and it hangs over any predictive enterprise. In fact, the world appears more akin to stock prices, which wander into the future with a jittery, random walk, than to the sun at dawn announcing the new day, which scientists beginning with Isaac Newton effectively explained as a mechanical process involving a relatively small number of (today) easily calculated variables. Today, prediction has grown more accurate.

Yet regularity is not necessarily the norm. Much of the world around us and in us is spontaneous and random. Many natural processes, from the weather to earthquakes to a lurking cancer to a novel coronavirus, are riddled with uncertainty. The mutational propensities of DNA may have an external cause—a blast of radiation, for instance—or an internal one, such as a genetic copying error. Tracing the origins of mutations may, in some cases, be done through modeling, but many remain opaque. There are also social realities driven by the decisions and actions of human agents, with their hidden currents, biases, complexities, and feedback loops.

Igor's amped-up dopamine gene might give him a propensity to take greater risks, but it does not ultimately determine his behavior or decision-making. Indeed, the knowledge that he can tolerate a high level of risk may make him more aware of its perils—nature *and* nurture. A prediction about behavior envisions an array of options or possibilities and the emergence of uncertainties, or risks, often with the goal of ruthlessly culling multiple possibilities down to the one that is the most likely. Both nature and nurture as well as their interactions are difficult to distill into just one future, though people insist on trying.

The First Games

For centuries after Newton, a kind of rigid determinism held sway over the scientific community, argues the Canadian philosopher Ian Hacking in his study of what he calls "the emergence of probability." Far into the twentieth century, the finest scientific minds believed that the appearance of randomness was simply our failure to fully comprehend natural laws; many still do. Determinists insisted that just one road from the past led into the future; freedom to choose was an illusion. In *The Taming of Chance*, Hacking quotes Hume, who, despite his congenital skepticism, characterized free will as a false belief: "Tis commonly allowed by philosophers that what the vulgar call chance is nothing but a secret and conceal'd cause." Even as late as the nineteenth century, the French mathematician Pierre-Simon Laplace, a key synthesizer of probability as a science, could imagine a deity-like intelligence behind the Newtonian universe "that, at a given instant, could comprehend all the forces by which nature is animated and the respective situation of the beings that make it up. . . . For such an intelligence, nothing would be uncertain, and the future, like the past, would be open to its eyes."

Yet, as Hacking suggests, chance was not so easily dismissed. The more deeply science probed the natural world, the more the random quality gave way to order—not the linear, mechanical order of Newton but something more complex. For example, the mechanisms underlying Charles Darwin's natural selection are still driven mostly by random genetic mutations, as was thought in the late nineteenth century, but there is now an appreciation of more complex factors, including genetic drift, location-specific radiation levels, transgenerational epigenetic inheritance, and host–microbial interactions, that are also genetically driven. Recent studies from Chris's lab and others have shown that humans are continually adapting and shifting immune cells relative to specific bacteria in their environment; even astronauts' T-cells shift to match the microbes on the walls of the International Space Station. The astronauts' bodies are adapting at the same time as the bacteria are evolving to resist detection, antibiotics, or containment, meaning that both systems (host and microbe) are required to understand the relevant biology.

In the twentieth century, physics revealed a quantum world that resisted determinism, consisting of particles defined by a probabilistic framework (for example, a wave function). Specifically, the stronger a prediction on the exact location of a particle, the weaker the measure will be for the momentum. Yet these wave equations can be analyzed statistically and measured with even greater precision than Newtonian mechanics ever could provide.

As Hacking points out, that resurgence of indeterminism—of chance—has produced a powerful paradox that could be the banner of modern science: "The more the indeterminism, the more the control."

But first, let us go back a few centuries. Although some of the first betting games and probability mapping can be found as early as ancient Egypt, the modern discipline of probability traces itself to a sixteenth-century Italian physician and mathematician, Girolamo Cardano, who enjoyed his time at the betting tables. In the 1520s, he wrote *Liber de ludo aleae*, or *The Book on Games of Chance*. Cardano appears to be among the first to wrestle with the "frequencies"—what we know as the odds—produced by a pair of dice tossed multiple times. He discerned an order beneath the random numbers of chance and took the first steps toward the quantification of risk and uncertainty.

Cardano was a Renaissance polymath—brilliant, but focused on a less than respectable game. By the 1660s, a remarkable group of European intellectuals had begun to explore the mathematics of probability beyond Cardano's games of chance: the deeply devout French mathematician Blaise Pascal, who applied probabilistic thinking to the decision to believe in God; his colleague, the mathematician Pierre de Fermat; as well as Antoine Arnauld, who led a group at the Parisian monastery Port-Royal, where Pascal had taken up residence, and who seems to have been the main author of an authoritative book on logic that contained four seminal chapters on probability. Meanwhile, a tight world of inquiring minds was making important contributions, often in response to real-world problems: the Dutch mathematician and astronomer Christiaan Huygens, the German mathematician and philosopher Gottfried Leibniz, and Johan de Witt, the Grand Pensionary of the Netherlands, who applied probabilistic techniques to an important mechanism for financing the Dutch government.

A key figure in this period was Jacob Bernoulli, who, as Peter Bernstein writes in *Against the Gods* (1996), was the first to consider linkages between probability and the quality of information, thus making probability applicable to a broad range of social statistics, such as patterns of births and deaths, disease, mental capacities, even the disposition of legal opinions. Stephen Stigler, in his authoritative history of statistics, calls Bernoulli "the father of the quantification of uncertainty." Bernoulli, part of a large, mathematically renowned Swiss family and a supporter of Leibniz's claim to have invented calculus, versus Newton, articulated what became known as the law of large numbers, which demonstrated that the more random occurrences there were—whether the tossing of dice or, in Bernoulli's favorite example, a known quantity of black or white pebbles in a jar—the closer you would get to a theoretical numerical figure that could be calculated beforehand: the abstract "likeliest" number. Bernoulli's law of large numbers did not dispel Hume's induction problem, but it did provide a tool for defining and reducing uncertainty. As Stigler writes, "What was new was Bernoulli's attempt to give formal treatment to the vague notion that the greater the accumulation of cases, the closer we are to certain knowledge about that proportion."

That brings us to Reverend Thomas Bayes, the English cleric who had an interest in probability and, before his death in 1761, wrote down a theorem that attempted to quantify the probability of an event taking place if the probability of its cause or causes—that is, its conditional or prior probability—was known. Like Bernoulli with his law of large numbers, Bayes did not directly refute Hume; he instead provided another method for quantifying probabilities by, as Bernstein notes, "mathematically blending new information into old information." Bayes's theorem took a while to become famous; he apparently never realized what he had wrought. Today, one of the key hypotheses about how the brain works focuses on the mind as a Bayesian inference mechanism—that is, a kind of prediction-error-minimization machine.

The Emergence of Statistics

As the study of probability grew more sophisticated, statistics also emerged more powerfully. Cities, particularly in Italy, had long gathered statistical data on their citizens, but now the eruption of tools from probability beginning in the sixteenth century drove data acquisition. Individuals and then states began to compile statistical materials—state numbers—starting in London in 1603 with weekly tallies of christenings and burials. John Graunt, a haberdasher with wide interests, used that trove of statistical material going back a century to calculate the first mortality and life-expectancy tables, essentially inventing demography and epidemiology. In the Netherlands, de Witt used statistical data on individual lives as part of his attempt to calculate a more accurate actuarial analysis of Dutch annuities. In 1700, Leibniz wrote up a plan for the Prussian state to gather and analyze statistical data. By the end of the eighteenth century, the decennial census was written into the US Constitution, the start of a now-vast collection of data that extends well beyond counting heads.

The essence of statistics is counting and measuring. As Francis Galton said, "Whenever you can, count." Galton, born in 1822, was the precocious son of an affluent English family; Charles Darwin was his half-cousin. Like Darwin, he had traveled extensively as a young man, particularly in the Middle East and Africa, and he was something of a scientific magpie. He believed in phrenology, which asserted that bumps on the skull were an indication of intellectual powers; he published Britain's first weather maps; and he pioneered the use of fingerprinting. But his real interests were heredity and inheritance—not terribly surprising in a near relative of the author of *On the Origin of Species*. Galton believed particularly in the ability to improve the human race through a kind of rational breeding program he called "eugenics."

He was also a devotee of statistics. In his history of eugenics, Yale University's Daniel Kevles notes that Galton, who was not a formally trained mathematician, brought statistics into biology "at a time when no biologist dealt with any part of his subject mathematically." In the early nineteenth century, the German mathematician Carl Friedrich Gauss had grown interested in the frequency of errors in astronomical observations. For a variety of reasons, astronomers rarely noted the exact same coordinates for heavenly objects. Gauss discovered that these errors, when plotted, formed a bell-shaped, or normal, curve (now called Gaussian). He recognized that a line bisecting the curve represented the theoretical mean, with more distant errors—outliers—falling in frequency the farther they were from the mean. In the mid-nineteenth century, the Belgian mathematician and astronomer Lambert Adolphe Jacques Quetelet made a splash by applying the Gaussian normal curve to a plethora of human measurements in an attempt to describe a theoretically average man.

Galton seized on the normal curve, but instead of attempting to define "normal" as "average," as Quetelet had, he focused on differences, particularly as they applied to heredity and questions about the distribution of talent or ability. In a series of experiments with sweet peas (one of his collaborators was Darwin), Galton recognized patterns in inheritance data suggesting that traits tended to revert (he used the more normative term *regress*) from one generation to the next. This finding challenged his belief that a better human stock could be rationally bred. Galton also failed in his attempt to pin certain physical traits to criminal tendencies. Yet he did recognize how physical traits, such as height, the length of an arm, or the size of a foot, appeared to be correlated in quantifiable ways. That could be explained by applying laws of probability to statistics, which allowed him to map a wide range of trait (phenotype) correlations.

The concepts of regression and correlation were major advances in probability and statistics and a credit to Galton, though he never succeeded in explaining the mechanism of inheritance or, in terms of biology, of regression and correlation. Experiments by the Moravian Augustinian friar Gregor Mendel, also with pea plants (which led to his speculations on the role of discrete, inheritable units—genes), took place during Galton's mid-nineteenth-century heyday but in near-total obscurity and failed to surface for decades. The rediscovery of the gene would be a seminal breakthrough in the twentieth century. However, despite Galton's failure to link physical characteristics to other qualities, his larger proposed program of human breeding, eugenics, flourished in America in part due to the broad belief that physical traits, or phenotypes, determined less-tangible qualities such as intelligence. Indeed, laws requiring the sterilization of so-called mental defectives, most often targeting minorities, occurred in 32 states in the early twentieth century. Eugenics found its most aggressive and awful embodiment in Nazi Germany's racial policies and concentration camps; though discredited today, it still lurks on the margins. Eugenics purported to be the scientific application of Darwinian evolution to human stock, and, for at least a century, it reveled in its scientific credentials, including its use of probability and statistics. It was at its heart predictive and in the end terribly wrong, both scientifically and morally, swallowed up by the risk of error that it had long ignored. It was also scientifically bankrupt, since it involved a prediction without a substrate. There wasn't a hint of either DNA or the genome at the time.

New Predictions from the Genome

In 2003, the Human Genome Project announced that an entire human genome had been sequenced for the first time. This was fantastic—except technically it wasn't the entire genome. The project sequenced some 92 percent of the genome, including 99.9 percent of the so-called euchromatin, which are loosely packed, uncoiled, and chemically active DNA sequences—the familiar protein-encoding DNA. But the initial sequencing didn't deal with the more obscure heterochromatin, chemically inactive DNA that's tightly packed in the chromosomes. (Heterochromatin is chromatin, the material of chromosomes, but it is found in the centromeres, which provide structural support as cells divide, and the telomeres, which cap and protect the ends of chromosomes with repetitive nucleotide sequences, TTAGGG-TTAGGG-TTAGGG, and shrink with the aging process.) In 2009, an update of the entire human genome still had 300 gaps; by 2022, however, that number had been reduced to zero.

Indeed, soon after the first human genome was published, reality set in—not just in scientific circles but on Wall Street and in the media. The elucidation of the genome, like so many scientific advances, served to point out how much we still didn't know, and it opened far more doors than it closed. Sequencing was a powerful tool to further explore the nuanced mechanisms of the genome within its cellular environment. The simplistic metaphor of the sequenced genome as a book of life was rapidly replaced by a far greater complexity of interacting networks, embodied in fields such as statistical bioscience, bioinformatics, and synthetic biology. Mastering the genetic code was not as simple as knowing the combination to a safe.

For example, of the 20,000 or so protein-coding genes in the human genome, most are exquisitely specific, used in a particular cell, in a specific manner, and at a particular time. To control these genes, complex epigenetic mechanisms are deployed that regulate how and when DNA is packed into structures—euchromatin (open chromatin) or heterochromatin (closed chromatin)—how proteins that compose the packaging are modified, and whether the DNA itself is methylated or modified in response to environmental or biological cues. At first, researchers thought there were only a handful of modifications to histones—the scaffolding that packages DNA—but now we know there are hundreds of ways to alter the chromatin. Also, DNA and RNA were first seen as stable; now we know that both DNA and RNA may undergo hundreds of modifications that can change the stability, accessibility, and function of specific genes. Moreover, from 2010 to 2020 an additional 10,000 genes were discovered in the human genome, showing that the fundamental task of counting the number of functional units of our human genetic heritage is still ongoing and changing.

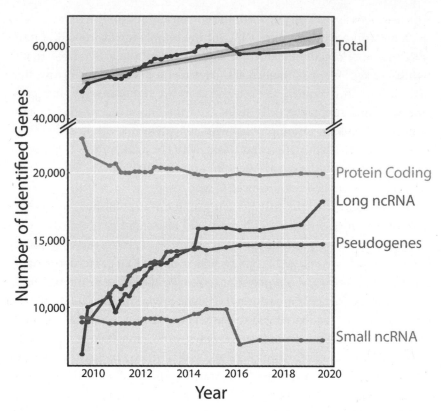

Figure 4.1

Continual discovery of human genes. The total number of genes (*top line*) found and annotated in the human genome continues to increase, coming mostly from the discovery of long, noncoding RNAs (ncRNAs) and pseudogenes. Almost all other gene categories (small ncRNA and protein coding) have been found.

There's a lesson here as we explore the changing relationship between prediction and risk; the key word is *change*. We have already touched on the exponential explosion of data and computing power. The innovative use of those two factors, fueled by several centuries of scientific research and technological innovation, enabled the success of the Human Genome Project. This means that the landscape we are exploring will not remain the same; it is changing even as we make our way through it. Indeed, it is changing *because* we are making our way through it. Similarly, as we sequence more DNA from other species and map the genetic dynamics of the world around us, we can readily see and map how the world is evolving and mutations are emerging every single day. Since life is always evolving and mutations are emerging every minute, any genetic catalog of life on Earth (or any planet) will be incomplete the moment it is finished.

Similarly, the number of algorithms and databases for analyzing and characterizing genomic data has increased at an exponential pace. This means that older data can and must be reanalyzed with (likely) improved and more comprehensive methods, and also that any database is also incomplete the moment it is indexed and loaded. The garden of biology and data is always growing and changing, and it must always be tended.

Genetic Risks

A few years after Chris gave Igor his genomic analysis, he asked Igor what he remembered about the process. "I thought a lot of the traits were accurate," Igor said. "But I remember thinking it was highly dependent on the algorithm you ran. There could be a lot of variability. After you updated the algorithm, I became *more* Jewish. A surprise was that I was 5 percent African."

A more recent algorithm and analysis of Igor's DNA knocked that down to 2 percent, but if you go back far enough, everyone has a little bit of Africa in them; that's where *Homo sapiens* came from. The models are always being updated, and a lot hinges on the data. It really depends on what's called anchor points in gene pools: sequences that mark the boundaries of geographics with a specific frequency of a gene—an allele. Just as various places in the world have unique genetic flora and fauna, the same is true of unique genetic variants that define human traits, pooled in various places around the world. The gene pool of an indigenous population tucked away behind remote mountains will remain hidden. We can see only the universe of data that we have access to; if certain data don't exist in a database, then certain human ancestors don't exist to that algorithm—a genomic version of the old philosophical puzzler about a tree falling in the forest. The algorithm might describe a 100 percent aboriginal Australian as Asian because there is a plethora of data on Asians but very little on aboriginal Australians. We're only as good as the models. The models define measurement and accuracy, and there's always a risk of missing data.

There are many perspectives on risk. There is the determinist view: if we know more, if we have more and better algorithms and more and better data—which are coming—we will be able to quantify risk and master it. In short, we could make risk go away. This is the Humean view. But some aspects of nature and life and even the structure of atoms themselves remain stubbornly probabilistic at best. How do we reduce the realm of risk to predict more accurately? How do we separate out quantifiable risk and deal with unquantifiable uncertainty? And what happens to risk as accurate prediction narrows it further and further?

In genetics, risk has already changed a great deal. The three billion bases of the human genetic code represent a lot of biology, and we already see extensive variation in the human stock. When we do a genomic analysis, we look for specific variants that we believe define a gene pool. A common tool of genomic analyses is a so-called confusion matrix, which is often used in fields such as diagnostic testing to compare the accuracy of a prediction of the spread of a disease with actual cases: the number of true positives and true negatives as well as of false positives and false negatives from which you can calculate accuracy and precision as probabilities. The Genome Analysis Toolkit (GATK) from the Broad Institute uses those methods to improve the discovery of genetic variants. Even though the GATK is one of the most widely used programs in genetics, it undergoes continual updating, which means laboratories must stay updated as well.

These updates are not just academic—there is a concrete legal risk if a lab doesn't stay updated. A key case around this issue started in 2005, when a four-month-old boy in South Carolina began to experience epileptic-like seizures. His physician ordered a genetic test and sent the DNA sample to a lab. The analysis revealed a mutation in the SCN1A gene, which provides instructions for creating sodium channels, but described it as "a variant of unknown significance." Doctors treated the infant based on that information; he died in 2008. But the mutation turned out to be not an unknown variant but one associated with a condition known as Dravet syndrome, which is rare, catastrophic, and characterized by severe seizures, often induced by fevers. Although there's no known treatment for Dravet, the regimen the doctors chose may have contributed to the child's death. The mother sued the diagnostic company, Quest, in federal court some eight years later for wrongful death, claiming it had failed to update its database. The case wandered through various South Carolina courts, including a stop at the state Supreme Court, which decided a narrow, if controversial, issue:

a diagnostics company is a "health care provider," like a hospital, and can take advantage of the state's relatively lenient malpractice procedures.

Having the most recent data is important—sometimes life-and-death important. Dravet syndrome is one of more than 25,000 diseases caused by defects in the genomic machinery, according to the Online Mendelian Inheritance in Man database. Some diseases are well known: Down syndrome, the most common chromosomal disorder, occurs when the body produces three rather than the usual two copies of chromosome 21. The risk of a Down syndrome baby, who may experience a wide variety of developmental delays, increases with the mother's age. Many mutations simply do not manifest a risk during a person's life, but others can lead to rare and sometimes lethal disorders. Although we have gotten much better at identifying and diagnosing these conditions, our ability to alleviate or cure the worst manifestations improves more slowly.

In fact, we all carry various alleles that either add to or subtract from our risks for diseases. This picture is not black-and-white, as it is for some genetic mutations; rather it is streaky and gray. In Igor's case, his genome showed he carries alleles that give him a higher risk of obesity, Type 2 diabetes, and coronary artery disease. He has some variants that could also give him a higher risk for atrial fibrillation, which can lead to a stroke. This indicates he should eat a heart-healthy diet and drink alcohol only in moderation to lower this risk. Some of his risks are 1.3 or 1.6 times higher than the baseline. Igor also showed mutations in some genes that could be important for what he has passed on to his children (such as breast cancer alleles in BRCA1) or could face later in his own life, such as prostate cancer, lung cancer, or age-related macular degeneration.

Then there's Igor's epigenetic age, a measure of how DNA's regulation changes over time. In all of our cells, the epigenetic state, measured by the degree of DNA methylation, slowly drifts from our primordial first cell. Just as sands move through an hourglass, DNA methylation marks in our blood and other tissues slowly drift over time, and they can be readily used to predict both chronological age—the number of times a person has circled the sun, measured by the candles on his birthday cakes—and biological age, a compilation of all the genetic and environmental factors in someone's biology. The variegated stresses, insults, and damage to DNA as well as frequency of exercise, eating habits, and general health contribute to the calculation of epigenetic age and indicate if someone is aging more quickly or more slowly than normal.

The good news for Igor is that his epigenetic age is lower than his chronological age. Being in shape helps. Igor was 51.3 chronological years old at the time, but when we tested his blood, whole blood, and purified T-cells, each had a different epigenetic age: 48.2, 48.5, and 49.3 years old, respectively. This shows that even very specific cell types, such as T-cells and other blood cells, can have slightly different (albeit consistently low) epigenetic ages.

There are ongoing efforts to better understand what drives these changes and how we can continue to further slow our biological ages. So far, evidence has shown that vegetarians have a lower epigenetic age than omnivores and meat eaters, that those who exercise are epigenetically younger, and that obesity drives one's epigenetic age higher. However, recent work has also shown that different tissues and organs have their own pace of aging, meaning that your heart could be biologically younger than your liver. Drinking excessively, for example, can damage the liver and age it more quickly. Smoking ages the lungs. In general, the more molecular insults you sling at any tissue, the more quickly it ages.

All these variables notwithstanding, epigenetic metrics for aging and DNA methylation are quite robust. Since some of the first work on epigenetic drift in the 1960s showed that such changes occur, follow-up work in 2005 by the Spanish genomics researcher Mario Fraga and other scientists such as Sven Bocklandt indicated, based on DNA methylation, that identical twins slowly become less identical over time. Thus, it has become clear that the epigenetic clock ticks away in each of us as the years pass. Subsequent work by the biostatistician Gregory Hannum at the University of California, San Diego, in 2012 and Steve Horvath at UCLA in 2013 established the key markers that broadly work across many tissues and sample types. This got the error estimate down to three to five years, even without knowing anything beyond epigenetic factors about the person, such as lifestyle (e.g., smoker, drinker, athlete, vegetarian). This also opens up the work for forensic purposes because any DNA left behind at the scene of a crime can roughly reveal the age of the person from whom it came. We delve into that subject more deeply in chapter 7, which may permanently change the way you think about crime and how we track it.

However, for these genetic, epigenetic, and molecular risks, it's important to note that higher risk is not necessarily a *significantly* larger danger.

For example, most of these disease risks are 30 to 60 percent higher, which might sound like a lot, but it has to be understood in the context of over-all risk. If you have a 1 in 10,000 chance of a danger in your life, and that chance goes up by 60 percent, then you have a 1.6 in 10,000 chance of that bad event happening. And all risk can be tempered with a bit of luck.

Of course, the risk of every prediction is different, and the definition of success will often vary dramatically from one enterprise to another. Broadly speaking, a financial trade that works out a little more than 50 percent of the time is going to be quite successful over time if that performance continues, which is difficult in markets prone to imitation. In medicine, the stakes are higher: lives are in the balance. Therefore, the demands on predictive accuracy in medicine—on accurate diagnosis and effective therapies—are higher. Generally, in medicine you have to be right at least 80 to 85 percent of the time, or people don't take you seriously. You wouldn't want to tell someone, "Well, you know, you might die, or you might not; I'll just flip a coin."

In medicine, the very definition of failure comes from Hippocrates: do no harm. A therapy that has no effect may not be worth the time and money to prescribe it, but at least it doesn't inflict injury. Defining success in treatment, however, can be difficult, an exercise in relativism, and may not be possible until decades have passed. There may be no immediate benefit or incentive, and probabilities can be complex. If there are no cures, then a 15 percent chance of success for a new therapy looks pretty good. Neuroblastomas (cancers in certain types of nerve cells) are very difficult to treat, but an improvement in diagnosis or treatment might mean a patient could get 11 instead of 10 months to live, and that extra month could mean a great deal to that person and their family and friends.

We wrestle with a final nuance in the chapters ahead: understanding how increasing predictive accuracy affects risk. Again, we must deal with moral hazard, the tendency of people to take risks or abandon prudence if they believe there is a safety net beneath them. A genomic analysis may convince some that they can eat and drink to excess because they appear not to have a predisposition to coronary disease or liver cancer.

How do we begin to take all these data, ever growing, and all those complexities and ambiguities into consideration? The answer is the increasingly sophisticated marriage of software and hardware that falls under the rubric of machine learning. To identify a mutation—say, the BRCA1 and BRCA2 mutations that can elevate the risk of breast and ovarian cancer—is, relatively speaking, straightforward, though in this case it's a risk factor, not an inevitability, and what you can do with that information is often a challenge. But, given the ease of DNA sequencing and analysis, such mutations are among the most straightforward signals of a future problem, certainly compared with the vast number of things that can go awry in one's life. To detect, diagnose, and understand the rapid changes in biology requires far more sophisticated tools, but many of these exciting new tools have arrived. That's where we'll turn next.

5   New Tools of Prediction

Cancer is one of medicine's most difficult problems of prediction. Cancer has many faces and can proliferate for many reasons in almost any cell in the body. Cancer is dynamic, like markets, like evolution. Accordingly, some cancers respond at first to a therapy, only to reemerge, shrugging off treatments that once worked to suppress or kill them and too often striking more lethally than before, like bacteria developing resistance to antibiotics or viruses to vaccines. Cancer features an amped-up engine of genetic mutations and epigenetic changes that helps it elude or resist the orthodox set of therapeutic options: surgery, chemotherapy, radiation, and, more recently, immunotherapy. Surgery may fail to remove all the cancerous cells. Chemotherapy and radiation may ravage both healthy and cancer cells alike and may drive the development of resistant cells, which, after a period of seeming disease regression, powerfully reemerge. The best option may be to detect cancer—or, better yet, the possibility of cancer—as early as possible, *before* carcinogenic outcomes. Of course, prediction isn't much use if something can't be done in response to it. Increasingly, after decades of trial and error, new, multicancer blood tests are beginning to roll out; even the first approved gene-therapy treatments—"repairing" mutational errors in the genome—have begun to appear.

In short, like the plethora of algorithms that Igor uses in the markets, medicine is rapidly building an armamentarium of increasingly predictive tools and models, from genomic analysis to sophisticated epidemiological models and data sets to machine learning, expanding far beyond rapid diagnostics, and now aiding drug discovery and continual monitoring for the first sign of any cancer.

Chris recently experienced this in his own family. In December 2018, something was wrong with his uncle, Ben, in Wisconsin. His gregarious uncle was suddenly tired all the time. He had a persistent cough, though he was not a smoker, and he would suddenly feel winded after climbing a flight of stairs. When Ben went in for a checkup, the family figured he had a case of the flu. Instead, the doctors told him he had lung cancer.

His doctors initially planned for a course of aggressive chemo and radiation. Then more data came back: Ben went from a diagnosis of one spot on a lung to several more in his lymph nodes. Then came even worse news; there were 18 "mets," or metastasized cancer spots, in his brain. The cancer had spread. He was presented with some difficult choices. He might have only two to three months to live. He could undergo harsh treatment to gain more time, but with more pain, or he could forgo therapy, live less long, and pleasantly enjoy the few days he had left.

Speed was essential. When Ben went in for a tumor biopsy, which would allow for analysis of the tumor DNA and possibly find clues to more effective treatment, the doctors failed to extract enough clean material. However, new treatments had recently emerged, driven by applications that have been possible only in the past decade, long after the human genome was first sequenced. These applications include the ability to sequence tumor DNA without needing to gain access to the tumor itself. This is possible because tumors and other cells shed DNA fragments known as cell-free DNA (cfDNA), which float through the bloodstream; these snippets of DNA can reveal significant clues about their tumor of origin. As a tumor grows, it releases more cells, and thus more DNA fragments. The fragments also contain signs of DNA methylation known as epigenetic marks, specific to each part of the body. Because each tissue has a specific methylation signature and tumors develop from tissue, the DNA methylation profile can show *both* tumor mutations and epigenetic clues of a DNA fragment's origin.

Chris suggested using the new cfDNA test to track mutations in his uncle's cancer. Like a hook on a fishing rod, the test captures molecules in the blood that match specific genes. Within two weeks, Ben's Wisconsin medical team and Chris's team at Weill Cornell in New York had sequenced and analyzed hundreds of genes from the tumors without having to extract them from the tumor. This genetics-from-a-distance test allowed the team to find a mutation in a gene called EGFR, which codes for a receptor protein in the cell membrane that responds to a growth initiator called "epidermal growth factor." The bad news was that this genetic aberration may have contributed to Ben's cells growing uncontrollably. The good news was that there was a drug that could target, or block, the receptor; Ben started on the drug immediately.

After eight months, most of Uncle Ben's tumors had disappeared, and the remaining spots had shrunk. The therapy worked, at least for a time, because precision medicine allowed his doctors to see aspects of biology that in the past were hidden. Ben returned to his normal life, including vacations and holidays with his family. Meanwhile, his doctors watched him carefully for any sign of recurrence.

Whole-Body Molecular Scan

Cancer is not the only area in medicine where cfDNA analysis can illuminate physiology. CfDNA has been used since the early 2010s to safely reveal information not only about one patient, but actually about multiple patients in one, such as a pregnant mother carrying a fetus. In 2014, Eric Topol, a Scripps Research Institute professor of molecular medicine, published a paper titled "Individualized Medicine from Prewomb to Tomb," describing how the nucleic acids of DNA and RNA provide data on the risk of specific diseases, miscarriage, or cancer for an individual even *before* birth. This kind of predictive medicine can have remarkable repercussions.

Historically, it has been difficult to gauge the health of babies before they are born, and it was dangerous to even attempt to gain genetic information from them. In the 1960s, amniocentesis was developed to collect chromosomal information about a fetus, usually to try to diagnose Down syndrome babies. The test involves taking a small amount of fluid from the amniotic sac and analyzing the fetal DNA, which then can be profiled for genetic abnormalities. Originally, the syringe used in the procedure was "blind," inserted according to an estimate of the best depth in the mother's swollen belly. Then in 1972, amniocentesis began to be guided by ultrasound to reduce the risk of miscarriage or fetal injury. Ultrasound then improved enough to visualize changes in the fetus and to detect problems without the needle.

But even with the invention of ultrasound mapping technologies, the visual cues (for example, gross anatomical malformations or nuchal translucency, a collection of fluid behind the neck in a fetus) did not necessarily reveal underlying genetic defects. In 1983, the Italian biologists Bruno Brambati and Giuseppe Simoni developed chorionic villus sampling (CVS), which tested placental tissue for genetic abnormalities. By the mid-1980s, physicians had three options for testing the health of babies before birth: ultrasound, amniocentesis, and CVS.

But there were risks. With both CVS and amniocentesis, there was a 1 to 2 percent risk of miscarriage. For many expecting parents, that risk was simply too high. Even with technical improvements in the early 2000s, the danger hung around a 1 percent chance of miscarriage, and parents—especially parents with increased risk of genetic abnormalities—faced the difficult decision of hoping for the best with a 5 to 10 percent chance of a genetic abnormality or taking a 1 percent chance of losing their child. Especially for women who had a hard time getting pregnant, this choice was between two awful options.

But all that changed in 2011 and 2012 with the emergence cfDNA profiling. With this new technology, the amniotic sac does not need to be punctured; pregnant women only have to have blood drawn from their arm—peripheral blood that contains cell-free fetal DNA (cffDNA)—and then have the plasma (the clear part of blood in a tube) separated from the cells, and the cfDNA can be sequenced. The mother's genetic profile can then be compared with the fetus's cfDNA, and traces of Down syndrome or other trisomy events that involve defects on chromosomes 18 and 13 can be mapped.

From the first proof-of-principle experiments in the early 2010s, it was clear that this technique had no greater risk than taking a normal blood sample, and researchers and clinicians quickly began testing variations of the sequencing and genome-profiling methods. It became evident that the sensitivity, specificity, and accuracy of the method were just as good, if not better, than the CVS and amniocentesis methods of prior decades. This led to large-scale clinical trials around the world that validated the results. By 2015, a flurry of new biotech companies that could perform cfDNA profiling for cell-free fetal DNA had launched. In 2010, 99 percent of fetal testing was by CVS and amniocentesis but only 1 percent by cfDNA, but by 2022 the numbers had flipped: CVS and amniocentesis now represent less than 1 percent of prenatal testing in the US market, and the rest is done with noninvasive, cfDNA prenatal testing. Cell-free analysis has proved to be an entirely new predictive medical procedure, with extraordinary power to sequence the genome of a baby before it is born.

Before birth, the amount of fetal DNA floating in a pregnant woman's bloodstream can be as high as 30 percent of her total DNA, and it steadily

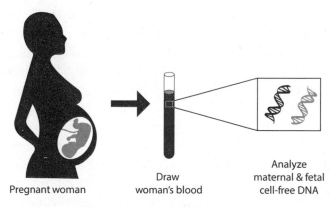

Figure 5.1
Origins of cell-free DNA for noninvasive fetal testing. The mother's blood, containing both her DNA (black) and placental DNA from the fetus (gray), the latter of which contributes to the cell-free DNA found in blood, can be used for noninvasive prenatal testing (NIPT).

increases with each month as gestation progresses. Yet within two to three hours of birth, fetal DNA disappears from the mother's bloodstream. This indicates that there is a continual stream of fetal DNA being splashed into the mother's bloodstream during pregnancy, representing a targeted loss of cells (known as apoptosis), a continual shedding of dead cells, and ejections of DNA by the fetus (figure 5.1).

CfDNA has also proved to be a key predictive tool in nonfetal settings. Another amazing predictive power of cfDNA is in heart transplants, where the organ recipient suddenly has a large mass of cells and the DNA of another human (the donor heart) inside his or her body. A study led by Cornell University biomedical engineer Iwijn De Vlaminck and Stanford University professor Stephen Quake in 2014 showed that cfDNA can predict that a heart transplant is being rejected by the body before doctors can detect the symptoms. As the host's immune cells attack the foreign organ, the organ's dying cells boost the amount of foreign DNA in the bloodstream, which serves as a measure of the rate and scale of degradation. This same concept has been applied to kidney, liver, and lung transplants.

In fact, cfDNA can act as a predictive molecular whole-body scan that can discern which tissues are dying. Sequenced cfDNA in urine has been used to discern if kidneys are being damaged from infection; the presence of nonhuman cfDNA from, say, microbes can be used to detect and diagnose the source of infection. This method of cfDNA urine sequencing is a simple, noninvasive method to profile possible low-grade infections and a way to examine the damaged states of tissues and cells around the body. Every fragment of DNA has its own origin story and possibly an origin from someone or something else (a fetus, a donor organ, or a new bacteria); we just have to listen.

Even tiny mitochondria—semiautonomous organelles in cells that generate energy inside of our own cells—turn out to be an early-warning system for stress. Mitochondria have their own DNA, called mtDNA, separate from human cellular DNA. MtDNA levels spike in the blood of people giving speeches, for instance. The Swedish psychoneuroendocrinologist Daniel Lindqvist and his colleagues showed elevated mtDNA in the blood plasma of people who had recently attempted suicide. And mtDNA peculiarities even occur in space, popping up in the NASA Twins Study, a large project led by the Mason lab that analyzed the genetic changes of identical twins who were both astronauts: Scott Kelly, who spent a year in the space station circling Earth, and Mark, who remained earthbound (and is now a US senator from Arizona, which is a whole different world). During his first week in space and again toward the end of his mission, Scott showed large increases in the amount of mtDNA in his blood, indicating significant immune stress from radiation, fluid shifts, and environmental changes. The fluid shift, for instance, involved three liters of volume that normally reside in the lower body suddenly moving into the upper torso and head during space flight, making the body react in strange ways, including spikes of mtDNA in the blood. This mtDNA spike, or lack thereof, may represent a way to gauge the health and stress levels of future astronauts.

Techniques such as cfDNA and circulating mtDNA combine personalization and predictability in precision medicine. These tools are a more sophisticated application of the genomic analysis discussed in the previous chapter. By using genomic tools to look deep into the complex biology of an individual patient, physicians can make an informed guess—a prediction—about what might be driving a disease and get clues as to how to stop it. This individual approach contrasts with methods we examine later in this chapter that analyze the effects of diseases such as smallpox and COVID-19 on broad populations. But the two endeavors are linked. These epidemiological techniques are predictive and involve model building, algorithms, and lots of data and computer power. Both perspectives, micro and macro, involve predictive strategies. And as we shall see, they echo and amplify similar techniques in many other fields.

Disease as a Math Problem

The emergence of the novel coronavirus SARS-CoV-2 in December 2019 posed significant and scary challenges. Although the genome of the virus was quickly sequenced and released in January 2020, allowing researchers to begin work on vaccines and track the cascade of new strains that emerged, many of the early guesses about its nature had to be quickly revised. Its origins and mutation rate were widely debated; how it spread and when were unclear; why some people were seemingly immune but others dangerously vulnerable was a mystery; the infectious potential of asymptomatic cases was unknown; and what kind of immunity the virus left behind was undetermined. This collection of unknowns spawned mixed messaging and sudden shifts in policy on matters such as wearing masks; it also opened up claims for untested "cures." Still, beneath all that, COVID-19's dynamics in the population resembled those of other pathogenic diseases. We have models, which will only improve as empirical knowledge of the virus and its variants deepen. And there are tools that crack the door open not only to predicting the course of the pandemic but also to finding out the wide range of ways the pandemic alters broader human behavior.

But long before COVID-19, there was smallpox, which arguably brought us the first marriage of data and disease as well as the first models of how a disease determines life expectancy and how it spreads. Scientists first believed that smallpox might have emerged thousands of years ago, based mostly on evidence of facial scarring in Egyptian mummies, descriptions that appear in fourth-century China and seventh-century India, and indications that the virus may have originated in an African rodent, the naked-soled gerbil. More recently, after discovering and sequencing the DNA of the actual virus found in a seventeenth-century child mummy buried in a Lithuanian church, some scientists argued that the infamous scourge developed as late as the fifteenth century—that the Egyptian mummy's facial scars came not from smallpox but from measles or chicken pox. However, even more recent genomic analysis of nucleotide mutations of viral DNA and its various ancestors suggests that smallpox likely emerged 3,000 to 4,000 years ago in Africa.

Regardless of origin, we know that the "speckled disease"—named for the smallpox pustules that cover the face and extremities of its victims—has been a major killer, striking rich and poor, kings and peasants alike, and is particularly deadly to children younger than ten (the opposite of COVID-19). In early-modern Europe, roughly one-third of those who contracted the disease died from it; cases and deaths mounted in the sixteenth and seventeenth centuries and peaked in the eighteenth century, when smallpox had a major demographic effect, with 15 million dying from it over a 25-year period and some 10 to 15 percent of all deaths attributable to it, 80 percent of them involving children. Like COVID-19, smallpox spread quickly. Sketchy accounts of rudimentary inoculations against what may have been smallpox suggest that the practice was invented in China in the tenth century, and by the sixteenth century had spread to India and to the Ottoman Empire and Europe.

The inoculation involved implanting a live smallpox virus, *variola virus*, just beneath the skin, producing an infection—although ideally not one severe enough to kill. The aristocrat and writer Lady Mary Wortley Montagu famously introduced the practice to Britain in 1721 after surviving an infection and witnessing inoculations in Turkey; she had both her children successfully inoculated. The possible payoff from inoculation was straightforward but grim: if you survived, you gained immunity against smallpox for life, but many died from the practice or were severely disfigured. Although some practitioners of inoculations claimed high survival records, the outcome often appeared to be arbitrary. But back then, much about medicine was mysterious.

In the decades before Edward Jenner developed the first "safe" smallpox vaccine in 1796 (derived from the nonlethal-to-humans cowpox virus), smallpox inoculations stirred a fierce debate in Europe that in many ways foreshadowed current antivaccine fears. Very little was known about the various strains of the virus, some of which produced more severe infections than others. Data were sketchy and imprecise, and misidentification rife. There was no safe way to know who was already immune. The temptation was to accept some deaths by inoculating children, but there were objections. Many believed inoculations were intrusions into God's right to decide who would become ill and who would live or die. And if a child was inoculated but then succumbed to the introduced virus, was that murder?

With the inoculation debate heating up, Daniel Bernoulli attempted to bring some certainty to that risk. Daniel was a son of Johann Bernoulli, a professor of mathematics in Groningen in the Netherlands, and a nephew of Jacob Bernoulli, who held the mathematics chair in Basel and discovered the law of large numbers. Clearly, math was in the family. Johann wanted Daniel to go into business, but he instead first studied medicine and then drifted back to the family trade, mathematics. After a stint in Peter the Great's Imperial Academy of Sciences in St. Petersburg—his older brother Nicolaus hired him—he returned to Basel, where he taught botany, physiology, and math.

Bernoulli's work in St. Petersburg focused on what are known as "utility functions" to help categorize risk. He distinguished between mathematical calculations of probability versus the subjective "gut" of individuals confronting risk. "The utility . . . is dependent on the particular circumstances of the person making the estimate," he wrote. "There is no reason to assume . . . the risks anticipated by each must be deemed equal in value." He then took this commonsensical, empirical notion one step further, arguing that for many who shoulder risk, the utility from an increase in wealth is inversely proportional to the quantity of goods previously possessed. Peter Bernstein offers his take on Bernoulli's work on utilities: "For the first time in history, Bernoulli is applying measurement to something that *cannot be counted* [Bernstein's italics]. He has acted as a go-between in the wedding of intuition and measurement." That work brought new quantification and predictive power to inference.

In his thirties, Bernoulli published a paper on a decision-theory puzzler known as the St. Petersburg paradox, which his brother Nicolaus had first laid out. The paradox involves a simple game of coin flipping in which a player gets paid every time the coin comes up heads; the payoff rises exponentially with the number of total flips: $2 on flip one, $4 on flip two, $8 on flip three. The game ends when the coin comes up tails and the player takes the pot. How much should a gambler pay to play the game? For a player, the game's rewards are theoretically infinite. But, as Bernoulli observed, gamblers resist putting up even a relatively small amount of money to take a chance on those infinite rewards. He concluded that there's an inverse relationship between wealth and utility: once you have enough wealth, the utility of getting more declines. Moreover, as later commentators on the St. Petersburg paradox noted, there's no casino in the world that could or would bankroll infinite gains.

With these odds in mind, Bernoulli was persuaded to tackle the inoculation problem in 1760. France was involved in an intense debate over smallpox inoculations. In a paper to the Académie des sciences in Paris, Bernoulli used the data from a study that Edmond Halley, Britain's Astronomer Royal and the discoverer of the comet that bears his name, had made of the population of Breslau, which was then part of the Hapsburg Empire (now in Poland). Halley's goal for the Breslau data, acquired with the help of Gottfried Leibniz, was to improve the actuarial basis of annuities, not to track smallpox deaths, but his study helped both. Bernoulli's work on smallpox thus was lauded by insurance actuaries, for whom life expectancy is a key metric, long before it was recognized as an epidemiological breakthrough.

Bernoulli broke the population into compartments: those who had not been infected and those who, having recovered, were immune. From that, he was able to extract some key insights: total deaths by all causes except smallpox; the rate that "susceptibles" were infected; the fatality rate from the infection, known as the force of infection; the probability for a newborn to survive to a given age. Bernoulli derived these insights through a series of differential equations. He concluded that fatalities from smallpox occurred in one of eight cases, or 12.5 percent, which allowed him in turn to calculate the chance of infection each year from birth to 25 years old.

From those calculations, he showed that the life expectancy for that group would rise from 11.1 years to 25.5—some 14 years—with inoculations. A study of Bernoulli by the Oxford historian Catriona Seth concluded that "the demonstration was striking on an immediate level as offering scientific support for inoculation, but also, more widely—though this possibly would have escaped the common man—for offering mathematics an interesting applied role in thinking of how contagion works."

Yet there were some flaws in Bernoulli's work, which was hardly predictive. Halley's data were suspect, and Bernoulli admitted that he could not get data on smallpox fatalities by age group. As a result, he used the assumption that all inoculated children would survive. The French philosopher Jean le Rond d'Alembert criticized him for treating the decision to inoculate as a cold, rational, quantifiable choice, which it certainly wasn't for any parent. This argument was often heard regarding COVID-19. But d'Alembert, a formidable mathematician in his own right, also suggested improvements to Bernoulli's calculations, arguing that the conception of risk changes dramatically with age. Bernoulli, annoyed by d'Alembert's criticism, insisted he would never again read the Frenchman's work, even though the two agreed on the overall merits of inoculations.

Making Room for Uncertainty

Although others made further improvements in Bernoulli's compartmental model, it mostly slumbered until the 1920s, when a renewed interest in mathematical disease modeling emerged. The Scottish physician and epidemiologist Anderson Gray McKendrick, who had worked as a public-health physician in Africa and as a lieutenant colonel in the Indian Medical Service, returned home in 1920 with his family after contracting a disease known as tropical sprue and took up a position as a lab supervisor at the Royal College of Physicians of Edinburgh. In 1925, McKendrick published a paper that included a Bernoulli-like compartmental approach to analyzing infectious diseases. More striking, McKendrick included nondeterministic random elements—that is, stochastic elements—in the model. In doing so, he anticipated work that was to come decades later in a number of other scientific fields, from hydrodynamics (a rushing stream is a stochastic system) to nuclear physics (streams of particles are stochastic) to telecommunications theory.

McKendrick's model incorporated a so-called Poisson distribution, named for Siméon Denis Poisson, a mid-nineteenth-century French mathematician and successor of Pierre-Simon Laplace. Poisson's work in probability focused mostly on criminal behavior and jury verdicts, and the formula at the heart of the Poisson distribution was a mere page in a much longer study he did in 1837, but, fortunately, another eminent French mathematician, Antoine-Augustin Cournot, recognized the power of the method and in 1843 explored it more thoroughly.

The Poisson distribution effectively took probability into the realm of chance. It is used to calculate the probability of events that occur randomly, such as a coin flip, in a given amount of time, area, or volume: customers arriving at a store, traffic passing on a street, the rate of radioactive decay, even search requests on the internet. It requires two assumptions. The first is that the events occur at a steady rate over time; you don't know when they will occur, but their rate can be calculated. The second is that events occur randomly and independently, not correlated in any way. The action of one customer, one search, or one particle is not linked to the actions of others.

By means of the Poisson distribution, McKendrick introduced mathematical tools to analyze aspects of disease spread that were not deterministic. After all, diseases spread not by viral forethought but by contagion, the random contact of individuals in a conducive environment. (Of course, the genetic evolution of new variants of pathogens is also mostly random, which makes pandemic prediction difficult.) Human interactions may be as random as stock prices. How many people capable of transmitting a disease will intersect with individuals susceptible to becoming infected at, say, a pool party? How will lockdowns, quarantines, or mask-wearing reshape the underlying math that drives an epidemic—the curves of contagion? What's the role of poverty or age in contagion or control? A stochastic process allows demography—births and deaths—to come into play. Births add to the most susceptible part of the population, while deaths subtract across all compartments. Both are independent and random in their timing.

Despite that insight, McKendrick decided to discontinue his initial stochastic models, turning instead to more conventional deterministic models using differential equations. Stochastic processes are useful in smaller groups, such as families, but deterministic methods work well with larger populations, where trends of mass action and movement apply. In 1927, McKendrick and the Scottish biochemist William Ogilvy Kermack, who had been blinded early in his career in a lab explosion, developed the so-called age-of-infection model, in which the infectivity of individuals depends on the time since they became infected. Kermack and McKendrick used three compartments in their model: susceptibles (S), infectives (I), and removed or recovered (R); therefore, the generic class of epidemic models that resulted from their work became known as SIR models. The pair envisioned individuals working through the three compartments over time. Importantly, the model allowed them to calculate a reproduction rate that determines whether the infection will die out, persist, or explode into an exponentially growing epidemic—that is, the model was predictive.

SIR models also underlie the modeled curves of the COVID-19 pandemic. The reproduction number (R_0)—the number of secondary infections caused by a single infected person—has also received a lot of attention. Generally, "flattening the curve" of infections requires a reproduction rate (R_0) of less than one. To achieve that, each new infected person must infect fewer than one other person—thus, the demands for social distancing, masks, and quarantines. Whenever R_0 is larger than one, it means the infection is spreading, sometimes explosively.

Kermack and McKendrick's SIR model became a kind of template for the far more complex epidemiological models that have proliferated since the 1950s, driven by the emergence of novel infectious diseases, by a far deeper understanding of how these infections spread, by increasing amounts of data, and by computers. These new models contain additional compartments: M (for newborns who temporarily receive disease immunities from their mothers) and E (exposed but not yet infected). Models such as MSEIR, SEIR, SEI, or SIRS are based on the flow of specific infections.

Beginning in the late 1940s with the work of the English statistician M. S. Bartlett, the models increasingly blended stochastic and deterministic elements, capturing far more complex social and biological realities and unifying the two sides of McKendrick's work. These ever more dynamic models have unfolded over time, driven by diverse variables. On computers, they can produce simulations of disease spread. They can be remarkably predictive if the data are good, the underlying understanding of the disease is relatively complete, and the ability to change social behavior is effective.

Alas, the human element remains an uncertainty. Some people reject the science, or at least what they think the science is saying. Others judge risk and utility differently—a problem of competing risks that Daniel Bernoulli identified centuries ago. Others resist being told what to do or have wildly different conceptions of self-interest. Many blame the models for failing to anticipate or accurately predict changes that may in fact be driven by subjective reactions to the models themselves. Success in flattening an infectious spread, for instance, may lead to overconfidence. Models do not determine or encompass reality; they're hostage to underlying assumptions and data. Alter those assumptions and data, and the prediction changes, just as the odds of an individual winning an election change as polls reflect changing voter sentiment.

In their various iterations, SIR models dominate infectious-disease epidemiology. When COVID-19 "infects" other aspects of behavior, understanding and predicting how the pandemic is reshaping myriad social forces must begin with that SIR foundation. For all the public-policy challenges during its spread, the COVID-19 pandemic helped to open broader predictive perspectives and possibilities that have been building for centuries.

A Layer Cake of Models

COVID-19 quickly changed the behavior of populations around the globe, albeit in different ways in different places as a result of varying public policies, cultures, economic systems, and levels of development. COVID-19 fits into the compartments of the basic SIR model; McKendrick's original notion of vectors moving from compartment to compartment of varying size broadly reflects the migration of the virus through a population. There is an underlying deterministic course to the disease: infections quickly spread through a population of susceptibles, producing recoveries and deaths. But there is also a stochastic element: community spread, the passage of infection throughout the population via people's interactions, is essentially random and, without effective therapies and vaccines, can be dampened only by reducing activity through quarantines or lockdowns or through the costly process of gaining "herd immunity"—that is, enough recovered cases that the reproduction rate, R_0, will fall below one and the virus will decline.

The COVID-19 coronavirus reshaped behavior beyond public-health measures. There was, of course, an enormous impact on the health-care system, schools, and lifestyles. With countries forced to lock down, to turn off their economies, many industries suffered. Others prospered: essential services such as groceries (Walmart and Kroger) and drugstores (CVS), food delivery (Grubhub and Uber Eats), online meeting tools (Zoom and WebEx), and digitally mediated retail delivery (Amazon.com). Broadly speaking, digital services thrived, from home cycling to videoconferencing to cloud computing. Restaurants and bars did not.

Industries and companies confronted a radically altered economic and commercial landscape. What products would be popular in the midst of a pandemic, and which would not? How would online sales compare with in-store sales? How would consumer trends such as pantry loading (the tendency in a crisis for consumers to hoard goods) and dips in sales (the result of earlier pantry loading) or the shift to meal takeout or delivery unfold in a world of social distancing? All of these dynamics evolved as internal and external shocks—COVID-19 spikes, supply constraints, the effect of therapies or vaccines—rippled through society, reshaping macroeconomic indices from declining disposable incomes to increasing household medical expenses to increasing (or decreasing) savings. Like the layers of regulation and coding that orchestrate the genome, the effects of COVID-19 resemble a layer cake of models: the SIR epidemic models overlaying macroeconomic models, atop increasingly microeconomic models for industries and companies, families, and individual consumers. Like the genome and all its genetic and epigenetic regulatory layers, this layered social and epidemiological structure is in constant flux, both deterministically and stochastically.

This is prediction on steroids—prediction applied extremely broadly. In fact, the pandemic may well have cracked open the door to a wider application of prediction. WorldQuant Predictive used SIR-like models to predict the spread of COVID-19, gathered as much data as possible from public sources and companies, satellites, and hospitals, and then built and tested diverse algorithms that attempted to predict relations among multiple variables. As we saw in chapter 3, these models can then be ensembled into metamodels or ensembles—fewer, perhaps, than the number of alphas WorldQuant assembles into investment portfolios, but the underlying idea is the same: the use of machine-learning systems to extract deeper patterns. The goal: predict risk, demand, infection, pricing, and value—and then act.

The possibilities are intriguing. Prognosticating consumer behavior isn't radically different from predicting biological activity—say, COVID-19 patients who are likely to develop dangerous cytokine storms, an immune *over*reaction, rather than immune suppression. An even more important question is whether these tools can be used to predict the molecules most likely to inhibit interactions between COVID-19 and a specific protein. In short, can this approach be used to significantly shorten the process of drug discovery?

So far, the answer is . . . sometimes yes. Companies have long sought to find molecules with chemical structures that bind to key receptors and block or amplify response; this search is often a long, laborious, and expensive process that turns on either mastering the three-dimensional structure of specific molecules, which fit into a protein receptor or enzyme, or conducting a brute-force mass screening of thousands of compounds. COVID-19 appears to be so efficient in transmission and infection because the spike proteins lining its outer surface bind to cellular receptors, allowing the virus to inject its RNA into a cell's replication machinery, suppress the immune response, and generate copies of itself by the millions in little time. (The coronavirus grows increasingly infectious as the spikes improve their efficiency through mutations.)

The key innovation is to break this mechanism down into a prediction problem—in this case, to find molecules from compounds that have already shown antiviral activity through mutations, including those from drugs used against HIV, Marburg, or Ebola, or those with efficacy against other coronaviruses, such as SARS, that might inhibit the ability of COVID-19 to enter and take over cells. Like any prediction problem, this breakdown requires several steps. First, it must define key targets, from protease receptors that allow entry through the cell membrane to the polymerase that enables viral replication within the cell. Second, it must get data. In the COVID case, the data came from a company called Chemical Abstracts Service (CAS), a division of the American Chemical Society, which for more than a century has quietly accumulated chemical data. CAS had a database of 50,000 known antivirals, providing WorldQuant Predictive with a treasure trove of observational data on bioactivity. Using that data, the company built algorithms and then ran them through machine learning, not seeking deep scientific answers but rather correlations—the stronger the correlation, the likelier the compound would have an effect.

This is empiricism, but with a difference. The process allowed WorldQuant Predictive to narrow the number of candidate compounds for testing. Of course, it's much too early to know whether this AI-driven screening program will be effective in broad drug discovery, where validation often takes years. But it does demonstrate the potential of these techniques. And if it works for COVID-19, there's every possibility that it could work for other pathogens, old and new, not to say more baffling diseases such as cancer.

Anticipating Resistance

Drug discovery is the Holy Grail of medicine, a metaphor, the biologist Richard Lewontin once warned, that molecular biology embraced in the run-up to the Human Genome Project, harkening back to "the most mystery-laden object of medieval Christianity." That said, patients want cures to their diseases—or at least control of them. Presumably, steady testing of cfDNA should provide some sign of new mutations or sets of mutations that could lead to resistance, whether through conventional chemotherapies and radiation, newer biological drugs, or targeted drugs, such as the one prescribed to Uncle Ben. Meanwhile, work on cancer resistance continues. In a computational biomedicine lab at Weill Cornell run by Olivier Elemento, a professor of physiology and biophysics and codirector of the WorldQuant Initiative for Prediction at Weill Cornell, scientists have pursued strategies that are conceptual cousins to WorldQuant Predictive's drug-discovery screening program.

A good example is a machine-learning algorithm developed by Neel Madhukar in the Elemento lab. Called the Bayesian ANalysis to Determine Drug Interaction Targets (BANDIT), the algorithm uses a diverse collection of data on drugs—some 20 million data points on six data types—to predict the most effective response from enzymes or receptors, both of which are often key to the proliferative tendencies of cancer cells. Employing data-cleaning methods used in finance, BANDIT aims to accelerate the traditional mass screening of compounds and the winnowing down of a large number of drugs to a select handful that can be tested. For instance, Elemento's group tackled a compound known as Onco201 that showed anti-cancer activity, though researchers did not know the compound's specific target. BANDIT concluded that it was targeting the dopamine receptor D4, which, as we saw in chapter 4, is one of the receptors that regulates risk taking and aggression. Because D4 is most commonly found in the brain, it's notoriously difficult to attack therapeutically.

In 2019, the Elemento lab published work with another lab led by Dan Landau that described a technique for "mapping" the capacity of tumor cells to develop resistance to specific drugs. This is a formidable computational task given the number of genes that can be involved in resistance—the paper focused on 10,000—and the growing number of anticancer drugs that are increasingly used in combination and with different dosages. The key is systematic sequencing and analyses of genomes in the lab, tracking mutations over time. The goal: to define likely resistance pathways and optimal drug regimens in advance, then narrow the field for clinical trials. One of the genes targeted in malignant lung cancer was EGFR, the growth receptor that had mutated in Uncle Ben's case. But as so often happens with cancer, the conclusions arrived at in this study are neither simple nor straightforward. Change is chaotic in cancer cells. Under pressure from drugs, some cells radically transform into stem cells. Others reestablish signaling mechanisms blocked by drugs. Two drugs in combination may find themselves blocked by genes that have no effect on each drug individually.

Those are just two forays against the phenomenon of cancer resistance. Work on the problem is being done in many medical centers and labs all over the world. The data pile up; the knowledge base deepens and broadens; and the models get better. The search for new signals and new correlations continues in every blood draw. Indeed, cfDNA can reveal which tissues in the body's cells are dying from COVID-19 just as well as it can map out which cells are dying from the targeted therapy in a cancer patient. But what is most striking is how computational and model-building techniques used to predict disease spread are also being employed in mapping pathways of resistance and, possibly, streamlining drug discovery and delivering new avenues to attack the cancer as it begins to escape. In the race against ever-evolving diseases such as cancer and viruses, predictive algorithms provide perhaps the best light for where to go next.

6   Mortality and Its Possibilities

In 2011, a professor of risk mathematics named Michael Powers bundled together some thoughts about risk and insurance and wrote a series of essays in a book that echoed Bernstein's *Against the Gods*. In *Acts of God and Man: Ruminations on Risk and Insurance*, Powers cuts to the chase on the first page of the first chapter, asserting that all matters of risk "flow from the same source: *the specter of mortality* [Powers's italics]. Like a serpent coiled around the trunk and branches of the Tree of Life, the risk of death squeezes at every aspect of human existence."

That's much more exciting, if a little grimmer, than the fine print of an insurance policy. But Powers identifies the core of insurance: the attempt to price what an actuary would call life's contingencies and what Powers characterizes as "life exposure"—the variable possibility of death that follows mortals as they move through life. He argues that both insurance and financial-risk management hinge upon life exposure. Insurance policies are designed to pay off relatively quickly, before "the policyholder's life terminates." And lenders must be compensated for the possibility that borrowers may, well, die (at least in the corporate form). Mortality hangs over so-called life tables, which try to capture the risk of living, breaking it down into one- or five-year slices and producing a curve of "mortality hazard rates" for each birthday (insurers and actuaries, Powers says, always assume that someone's death will occur on his or her birthday; this does not make them popular). Powers reviews the complexities of such an analysis, and, even with lots of data, insurers remain at some distance from the fate of individuals because of the underlying role played by health, class, and racial differences.

Powers looks to a future when science may be able to take greater control over the aging process—and with that, our Quantasaurus, with its data-driven reduction of risk, lumbers into the room. He writes, "As we approach that privileged time, the notion of risk inevitably will undergo dramatic change and perhaps even disappear from the human vocabulary." In fact, some of this change, Powers notes, has already begun. Modern medicine has helped flatten (though not eliminate) both the once-robust hazards of being born and of being very young as well as the similar rising curve that occurs with advancing age. Powers posits a time when the entire curve has been flattened to such an extent that every age period has the same hazard of dying in the next year. Average life expectancy, not surprisingly, will rise; although death will still occur from accidents and disease and suicide, "the rate of death will be independent of age." Although there are no guarantees of not dying young, Powers writes, individuals will not feel that they are approaching death—that time is running out, like the proverbial sands through the hourglass. Death will appear randomly, striking at any time. Powers mulls over how people might react in such a situation. Will the perception of death be more or less terrible? Will the death of a very old person be viewed as more or less a blow than the sudden demise of a younger person? In theory, it would be the same.

Powers's thought experiment is similar to what we are pursuing here: How do behavior and perceptions change as the ability to predict advances exponentially and risk proportionately retreats? Such change shouldn't be a surprise; the insurance industry and actuarial science have long sought to more accurately predict outcomes and fix prices on risk. Actuarial science has grown increasingly sophisticated and data rich, but it always faces the inherent limitations posed by any predictive enterprise because of its Humean uncertainties: it uses data from the past to provide glimpses of the future. Indeed, insurance did not become predictive until the science of statistics married the tools of probability in the seventeenth century. The notion of a flattened mortality hazard curve is a kind of prediction, based on data, that changes the nature of the risk. Insurance is thus arguably the world's oldest application of prediction and risk, one that is currently being transformed by many of the powerful tools that we have already described: exponentially growing data and computer power as well as a host of algorithmically based techniques. Like Powers's notion of a flattening hazard curve, the ultimate results may not be fully here yet, but the process has clearly begun.

A Response to Uncertainty

For a product that few people understand, insurance has been with us for a very long time. Its original uses appear to have been to grease the wheels of commerce, especially in trade. Reports from China as far back as the third millennium BCE describe efforts by traders to reduce the risk of loss by spreading goods across multiple ships. Babylonians developed a scheme in which merchants took out loans and paid an extra amount for a proviso that if a ship were lost, the loan would be canceled; it was thought an important enough innovation to end up in the Code of Hammurabi in 1750 BCE. The Achaemenid Empire in Persia in the third century BCE developed a form of life insurance. Greek and Roman traders bought insurance policies with premiums that fluctuated based on the risk; the Roman emperor Claudius personally insured shipments for the transport of corn. Benevolent societies arose to insure the life and health of their members— perhaps the first social welfare system.

Insurance slowly evolved to cope with a world rife with uncertainties. Thomas Hobbes's description of humankind's lives as "solitary, poor, nasty, brutish, and short" was true of his own seventeenth century and had been true since prehistory. (Ironically, just as Hobbes wrote those words, uncertainties were beginning to shrink.) Back then, mortality loomed large over humankind, from natural disasters to infectious diseases to eruptions of violence to lack of workplace safety. Lives and livelihoods were besieged by risk. Insurance would not prevent infectious diseases such as smallpox or various plagues or the ever-present threats of famine and pillaging, but it could make life better for the survivors. The demand for *assurance* drove the pricing of uncertainty and thus attempts at prediction, however crude.

Insurance was an attempt to reduce the harm for the survivors of events that could not be specifically predicted, even though it was reasonably certain that they would occur at some time—the maritime storm, the city fire, the sudden death from lightning or a burst aneurysm. The annuity, in contrast, flipped the equation: rather than be compensated for loss, the beneficiary received regular payments to support life until death. The state (historically the only enterprise with the resources to fund such a venture) sold the rights to participate in the annuity for a figure that was priced at a certain number of times the annual income that annuitants would receive in return. The state then invested the fund at a prevailing interest rate and paid the annuitants out of the proceeds until they died. The state got to use a large pot of money plus the tax proceeds on the payments. The annuity gave people a theoretically consistent, safe source of income.

In the Roman Empire, life annuities grew common, mostly as a means of passing along wealth between generations or of rewarding loyal retainers, though no one yet understood that pricing should vary with the age of the annuitant; most annuities were bought for small children, who, if they survived infancy, had a pretty decent chance of living a full life and maximizing their income. Annuities also inevitably posed tax questions: How do you tax an annuity, and how do you predict those tax revenues? Annuities benefited the living, but to finance them the state required some sense of mortality and life expectancy.

In his history of finance *Money Changes Everything* (2016), William Goetzmann calls the annuity contract "one of Europe's greatest contributions to humanity." Goetzmann contends that the ability to write a contract on a single life or a group of lives allowed citizens to shift "the risk of longevity or untimely death" from the family to the state. Annuities were an early form of social welfare or social insurance. The state could pool and diversify risks, but at a cost: the state rarely understood or had the tools to calculate its total risk exposure over longer periods of time. The answer came from probability and statistics, but those nascent fields were still gestating in the Roman era. Insolvency was a risk in any annuity scheme. Even today, Social Security, the most ubiquitous annuity in the United States, carries the continual risk that it will run out of money as the rate of population growth slows.

Controlling annuity risk requires an understanding of demographics. How large is the population? What is the average citizen's life expectancy at different ages? What is the balance of births and deaths or the effect of immigration and emigration? What role does infant mortality play? The rate of growth for humanity has changed many times over the centuries (figure 6.1), but it is now starting to flatline.

Enter a Roman lawyer, Domitius Ulpianus, more commonly known as Ulpian. Born in the Phoenician port of Tyre, not far from modern Beirut, Ulpian appears as a senior civil servant in the Roman bureaucracy in the third century CE, first under Emperor Septimius Severus, then, after the emperor's death in 211, under his sons, who engaged in a bloody succession struggle. In 222, young Alexander Severus, helped by his sister and mother, seized control, and Ulpian, who by then had codified and written commentaries on a vast swath of Roman law, became prefect of the Praetorian guard, the emperor's personal military wing. This was a powerful and precarious position. In 223, in Alexander's presence, some Praetorian guards murdered Ulpian during riots between Roman citizens and the guard.

Historically speaking, Ulpian is less known for his career in the civil service or his bloody end than for his extensive legal commentaries. Three centuries after his death, the Byzantine emperor Justinian ordered the compilation of Roman legal commentaries in what became known as Justinian's Digest. Some two-fifths of the Digest, one of the most influential works in Western legal history, is attributed to Ulpian.

Buried in the Digest is the earliest of life tables—Ulpian's life table.

There are complexities to Ulpian's attempt to grasp essential demographics for Roman life annuities. He likely did not assemble the life table himself, but he quite clearly used statistical material on Roman births and deaths that existed within the empire's tax bureaucracy. That compilation of statistical material is interesting in itself: the bureaucrats were counting heads and keeping records, although those records have never been found. The life table was included in the Digest in a roundabout way. Another near-contemporary jurist, Aemilius Macer, wrote of Ulpian's life table as part of Macer's own commentaries on tax policies from the reign of Augustus three centuries earlier. The table was not employed in the pricing of annuities, as it might be today, but instead used to forecast tax inflows from annuities depending on the age and life expectancy of the annuitant.

Despite all that, Ulpian's life table had a considerable afterlife, providing a rare illumination of some of the most intimate details of the Roman Empire: births and deaths. It would be cited well into the nineteenth century, when it was seen not just as a mirror of Roman society but as an anticipation of the modern actuarial table.

Life tables generally capture a specific demography and provide an anatomy of the inner dynamics of human populations. In a study of Ulpian's life table in 1982, the University of Michigan classics professor Bruce Frier outlined the essentials of historical demography. "The required data fall into three classes: first, the functions of mortality indicating how many males and females of varying ages will pass out of the population through death; second, the functions of fertility indicating how many new members

---

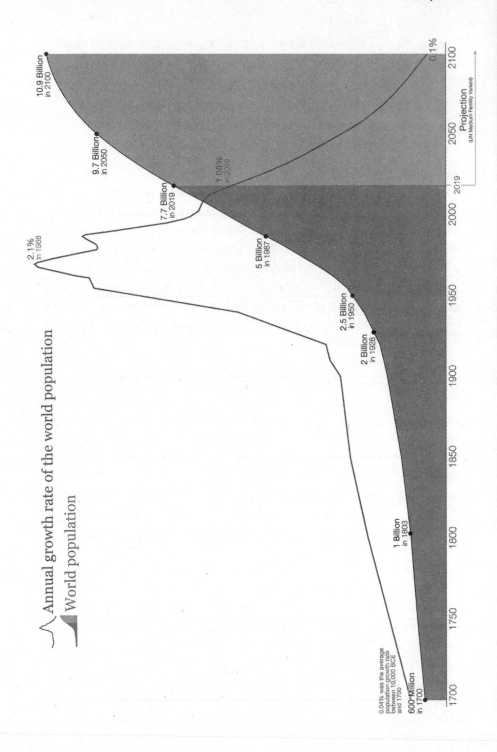

∿ Annual growth rate of the world population

◢ World population

2.1%
in 1968

1.08%
in 2019

0.1%

0.04% was the average
population growth rate
between 10,000 BCE
and 1700

600 Million
in 1700

1 Billion
in 1803

2 Billion
in 1928

2.5 Billion
in 1950

5 Billion
in 1987

7.7 Billion
in 2019

9.7 Billion
in 2050

10.9 Billion
in 2100

Projection
(UN Medium Fertility Variant)

1700   1750   1800   1850   1900   1950   2000   2019   2050   2100

will be born into the population; and third, the functions of migration indicating the transfer of members into and out of the population." Birth, death, and the movement of people—these, wrote Frier, are "universal and profoundly affecting themes."

Ulpian's life table had its deficiencies, as Frier makes clear. No one really knows what Ulpian considered to be the population base—some scholars have argued that it consisted of slaves or former slaves who received annuities after their owners died—or what assumptions and statistical methods, if any, lay behind the numbers. His notion of life expectancy was not the average number of years still left at a specific age but rather the median, or the number of years it would take until half of a given age group was dead. Perhaps most significantly, Ulpian's life table ignored the large number of children who died before the age of five, particularly from infectious diseases, which created enormous demographic pressures on families to produce more children. As Frier notes, the omission of children may simply have been practical: children didn't pay taxes. But it's also true that the tools to link mortality in age groups to cause-of-death patterns, "bringing together thereby demography and epidemiology," did not yet exist.

How well did Ulpian's life table reflect Roman reality? Given all the caveats, Frier thinks it did so reasonably well. He calculates that Ulpian's life table reflects a life expectancy at birth of between 19 and 23 years, driven by infant mortality of some 499 out of 1,000—not that different from the demographics of the Bronze Age.

Rebirth of the Life Table

Ulpian's life table stands out prominently because, oddly, there would be nothing like it for another 14 centuries. But by the time demography began

---

Figure 6.1
Rate of growth and world population, 1700–2100. The population growth of the world rapidly accelerated after 1900, but the rate of growth is now slowing and is expecting to be near flat by the year 2100. *Source*: Max Roser, Hannah Ritchie, and Esteban Ortiz-Ospina, "World Population Growth," Our World in Data, 2013, https://ourworldindata.org/world-population-growth (CC BY), based on data from History Database of the Global Environment (HYDE), the United Nations (UN), and the UN Population Division.

to systematically count things, in the mid-seventeenth century, many of the same dynamics behind Ulpian's life table had recurred: again, annuities played a central role, and, again, the state was a key player that needed data from life tables. Notably, the rebirth of these statistics came from two key sources: Johan de Witt's pioneering work on annuities in the Netherlands in the 1670s, and Edmond Halley's statistical breakdown of the population of a middling city in the Holy Roman Empire two decades later.

De Witt was an Ulpian-like figure who came to an Ulpian-like end. The son of a politician, he was a mathematician of some skill who became the grand pensionary of Holland—leader of the largest and richest state in the Netherlands—in 1653 at age 28. This made him a kind of prime minister of the Dutch Republic. He held the job for 19 years. This was Holland's Golden Age; having freed itself from Spain in 1648, the small country became a global trading and financial power of increasing prosperity. In 1654, it narrowly lost the first Anglo–Dutch War, then in 1667 it won the second; both were fought to control maritime trade. For decades, the Dutch had financed the government by selling annuities and perpetuities, bonds with no fixed maturity date. But as the wars began, the importance of annuities as a fundraising vehicle increased. By 1670, de Witt, who had read John Graunt's statistical study of London's birth and deaths, the *Bills of Mortality* (1662), was convinced that the government was pricing life annuities too cheaply at 14 times annual income. De Witt's study was the first rigorous analysis of annuities.

There were inevitable limitations to that analysis. For one thing, de Witt lacked solid data. He finally turned to a close friend, Johannes Hudde, the mayor of Amsterdam and another skilled mathematician, who had played a role in Newton's development of calculus and had helped design the curves of dikes. Hudde mined data from Dutch annuities going back to the late sixteenth century, but that did not provide the kind of robust birth-to-death data that was necessary for de Witt's analysis. So de Witt instead made a series of generalizations about infant mortality and the uniformity of mortality rates for every age level older than three and younger than 80; his life-expectancy chart resembled stairs—constant for a period, then a sudden leap, then constant again—and he and Hudde argued over this picture. As Goetzmann writes, de Witt "attacked this question with mathematics," marrying the calculation of the time value of money with the still-infant mathematics of probability. "De Witt recognized that the central challenge in the calculation of life annuity values was the probability of life expectancy—the need to estimate how long the cash flows would extend forward through time, on average, for annuitants of different ages." He sought, in other words, to quantify the very risk that annuitants "insured themselves against." He concluded that the annuities rate should be priced not at 14 times annual income but at 16 times, based on a prevailing 4 percent interest rate.

De Witt shared something else with Ulpian: a violent end. In 1672, a year after he completed his annuities analysis, France, Britain, and two German states went to war with the Netherlands. Tensions rose between the republican state led by the grand pensionary and the ambitious young Dutch king, William III of the House of Orange. De Witt, who was blamed for the war and for resisting the king, resigned after a series of Dutch defeats. In August, his brother, Cornelis, was imprisoned in the Hague on charges that he had threatened to murder the king. Johan, who had survived one assassination attempt, was lured to the prison with a fake letter. A mob dragged the brothers out of the jail and lynched them in the street. De Witt's annuities plan was never implemented.

The war drove the Netherlands into a deep financial crisis and ended inconclusively in 1678. King William III, whose father had died of smallpox eight days before his birth, by then had assumed control of the Dutch state. In 1688, he was offered the crown of Britain after the Glorious Revolution forced out the Catholic king Charles II. William brought to Britain many of the innovative financial techniques that the Dutch had employed throughout the war years, including the use of annuities as a source of government funding. In many ways, William and his wars nurtured the rise of London as a global center of finance.

Getting a Fix on Mortality

The second big advance in insurance came from Edmond Halley. Powers credits Halley with the first "comprehensive mortality table," noting how appropriate it was that, given the association of mortality with fate and destiny, the man "who made the first scientific prediction of a comet's appearance—long considered a portent of good or bad fortune—should also offer the first scientific analysis of the human life span."

The year was 1693. William III had become King William of Britain. The British government planned a significant increase in the issuance of life annuities, aimed at financing the Protestant William's continuing war with Catholic France. Halley grew interested in the same problem posed by annuities that had driven de Witt: pricing. Like Dutch annuities, the traditional British offerings were not age dependent; that is, the cost of an annuity for a five-year-old who might live to be 80 was the same as the cost of an annuity for a 50-year-old.

Halley, like de Witt, had a data problem. Current data (from Graunt's statistics) did not give ages at death or the number of people surviving at a given age, so Halley couldn't calculate life expectancy. He also lacked a stable sense of the population in a city with large immigrant flows. However, a Lutheran minister, Caspar Neumann, had gathered mortality statistics from 1687 to 1691 in the much smaller city of Breslau, and he had shared them with Leibniz, the court librarian of the Electors of Hanover. Liebniz passed them to Henry Justel, the royal librarian of England, who gave them to the Royal Society, a scientific organization where Halley worked as a clerk. (Significantly, Liebniz noted that Neumann's data effectively refuted the ancient concept of climacteric years, the belief that certain chronological ages saw more deaths than others and that deaths were influenced by the moon.)

Halley used the Breslau data to construct a life table, did some calculations, and published his analysis in the Royal Society's *Philosophical Transactions*. He noted that, based on the data, the British government was pricing its annuities far too cheaply at around six times annual income, particularly considering that pricing did not vary with age (echoing de Witt's conclusion). The Breslau data gave Halley the ability to vary his analysis by age, as modern actuarial analysis does. Using his table and an interest rate of 6 percent, Halley produced an annuity that was priced at 13.4 times annual income for a 10-year-old and at 7.6 times for a 60-year-old. When the government finally did sell the annuity, more than half of the 1,002 lives that subscribed were younger than 11 years, suggesting that investors knew age insensitivity was a bargain, albeit one that the government wouldn't realize for decades.

Halley's Breslau data found other uses. As we've seen, 40 years later Daniel Bernoulli used them to complete his analysis of the effect of smallpox inoculations on life expectancy that merged epidemiology, demography, and the relatively new sciences of probability and statistics. Halley's analysis was hardly perfect. He left out data on ages older than 85, and as with his handwritten astronomical calculations, he took shortcuts, fudging and smoothing some of the results. But the results were still powerful. In the decades ahead, Britain effectively built a mighty financial engine, centered in the City of London, around new institutions such as the Royal Society (founded in 1660), the Bank of England (1694), and the maritime insurance bought and sold at Edward Lloyd's coffeehouse on Lombard Street—a business that eventually became Lloyds of London. (The term *underwriters* came from investors willing to accept some liability for potential losses by writing their names under the line.) These institutions provided what the historian John Brewer calls "the sinews of power" that fueled Britain's aggressive global expansion. The conflation of mathematical reasoning and useful knowledge was embodied in "political arithmeticians" such as Graunt and Halley, notes Brewer. "The object of applying mathematical science to 'useful knowledge' was to make accurate decisions, to reduce elements of chance and caprice in what was perceived to be an unpredictable world. The risks and dangers associated with 'Meer Hazard and Chance' were contrasted with the security conferred by reason and computation." Brewer adds, however, that the confidence in mathematical probability, "stretched to a point of extreme credulity," also helped create a proliferation of popular life insurance schemes. Some were simply scams for raising money. Others, lacking the rigorous mathematical basis they touted, went out of business.

Beginning in the late eighteenth century, insurance products were increasingly offered outside of government. A key figure was the remarkable Richard Price: Unitarian minister; pamphleteer in support of the American and French Revolutions (and target for Edmund Burke in *Reflections on the Revolution in France* [1790]); friend of Benjamin Franklin, John Adams, and Thomas Jefferson; philosopher; mathematician; and literary executor for the statistician and Church of England minister Thomas Bayes. Price, in fact, not only presented Bayes's theorem on statistical inference at the Royal Society but also wrote an introduction that put philosophical flesh on Bayes's bare-bones observations. Price was also a formidable demographer and what historian Craig Turnbull called "a titan of actuarial thought" and "arguably the first actuary." He advised a number of early insurance companies, including the Society for Equitable Assurances on Lives and Survivorship, one of the pioneers of the new industry; his mortality tables from eighteenth-century Northampton and Norwich parish registers provided the actuarial foundations for the company's groundbreaking policy of pricing premiums by age. The Northampton tables quickly became canonical in Britain and the United States, despite Price's overestimation of mortality—later studies suggest he calculated life expectancy at 24 years when it now appears to have been around 30 and improving rapidly with declining infant mortality. As the historian Ian Hacking notes, Price's conservative table worked well for life insurance. Because the population was living longer, insurers could take longer to pay out while collecting premiums. The situation, however, was the opposite with annuities. "In 1808, the British government, hard pressed by war and inflation, decided to issue annuities 'soundly based' on Price's tables," wrote Hacking. "Hence it lost millions of pounds because people lived longer than is implied by any sound insurance table."

In 1771, Price published a book with the very long, if descriptive, title *Observations on Reversionary Payments; on Schemes for Providing Annuities for Widows, and for Persons in Old Age; on the Method of Calculating the Values of Assurances on Lives; and on the National Debt. To Which Are Added, Four Essays on Different Subjects in the Doctrine of Life-Annuities and Political Arithmetick.* One could simply call it *Age- and Risk-Adjusted Annuities.* In addition to publishing Price's mortality tables, *Observations* discussed how to adjust for immigration and the tendency to understate infant mortality. Price offered some generalizations learned at the Society for Equitable Assurances: (1) avoid adverse selection, such as writing too many policies for any specific risk; (2) how to deal with adverse variation in experience, such as war and plague; and (3) how to distribute the surplus that a well-run insurance scheme can generate. As Turnbull's history of actuarial science declares, Price was "inarguably one of the most important and influential actuaries in the profession's history."

By the nineteenth century, advances in probability and statistics embodied in life (or death) tables reshaped the nature and process of insurance and the underlying actuarial science. But as Price's *Observations* suggest, the great advances were organizational. The credit for much of the latter goes to an insurance entrepreneur named James Dodson, who not only took advice from Price but also launched what would become Equitable Life in the 1760s. (Dodson died in 1757, before a nervous government approved what Turnbull calls "a revolutionary life assurance design" in 1762.) Dodson's insight was how to effectively package the predictive conclusions from Price's table into an ongoing enterprise, in particular the development of age-specific pricing in so-called whole-of-life policies, which covered an entire life, as opposed to short-term policies. Dodson, in effect, married prediction and the quantification of risk and made insurance a mass product. He also developed an innovative financing mechanism for his new company, including two tiers of investors: participating investors, who bought policies and shares in the company, and buyers of only the insurance. The surplus from operations went to the participating investors.

Modern Models

With its mutual financial structure and whole-of-life policies, Dodson's Equitable became the model for many insurers for more than 200 years, up to the present. But today that model is changing with incredible speed. Behind the change are two factors we've visited again and again: (1) the sheer growth of data of all kinds from a densely woven cocoon of digital sensors held together by the internet, powered by the Cloud and rapidly advancing mobile, Wi-Fi, and automation technology; and (2) the full panoply of AI, which can handle enormous data sets and can learn. The mounting tide of data, particularly from the internet of things (IoT), offers increasingly sophisticated tools to predict and price individual outcomes, both in the long term and in the short. This is a long way from Equitable Life's whole-of-life strategy. For most of recorded human history, only a handful of risks mattered—essentially, property and life. Mortality may be the ultimate source of risk, as Powers says, but it has increasingly become surrounded by the underbrush of other risks. Coverage is disaggregating to cover smaller, more easily quantified risks: travel insurance, phone and phone battery insurance, flight-delay insurance, separate coverage for washers and dryers, vacation-package insurance, pet insurance, catastrophe insurance, insurance on insurance.

The data may come from wearables, electronic devices that can be worn and may monitor individual health metrics; IoT devices in your car (driving or self-driving); increasingly "smarter" homes; and secondary or proxy data. To take advantage of this trend, insurers will have to shift from the traditional practice of pricing risk by category, which included demography, income group, age, or sex. Risk will no longer be a purely group

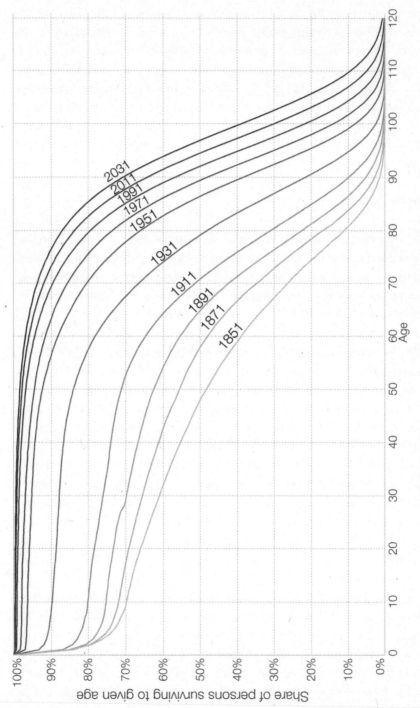

2031
2011
1991
1971
1951
1931
1911
1891
1871
1851

Age

Share of persons surviving to given age

Note: Life expectancy figures are not available for the UK before 1951; for long historical trends England and Wales data are used

phenomenon based on the past but a real-time feed of data from your car on your driving habits, activities in your house, posts on social media, and even fitness monitors on your body. Insurance companies now give discounts for adding devices to policyholders' cars ("safe driver" programs) and homes, and use these data to adjust rates. Much of these data flow through your phone and onto the internet. Insurance is now less about a single event such as a car accident, illness, or death and more about the constant flow of risk, which will wax and wane as you go through your day and your life, constantly altering and adjusting premiums.

In short, the prediction of risk is steadily moving toward real time, and these tools will be needed to move the maximum life expectancy beyond 122 (the current record for a human; see figure 6.2). This isn't science fiction. As a McKinsey & Co. report on insurance and AI from 2021 notes, "All the technologies required . . . already exist, and many are available to consumers. With the new wave of deep learning techniques, such as convolutional neural networks, artificial intelligence has the potential to live up to its promise of mimicking the perception, reasoning, learning, and problem solving of the human mind. In this evolution, insurance will shift from its current state of 'detect and repair' to 'predict and prevent.'"

Insurance is also being personalized, much like health care. No two customers are alike, and differences among them will only grow. Insurers will want that stream of data coming from your wearable or your regular genetics test. Gained a few pounds over the holidays? That information, along with your cholesterol level and blood pressure, could be transmitted to your insurer, which may contact you about a healthier diet or offer the number of a nearby gym while slightly increasing your premium. Lose a few pounds, keep up with regular checkups (supplemented by data from wearables), engage in increased exercise, and your health premiums may

---

Figure 6.2
Survivorship according to mortality rates experienced or projected in persons born 1851–2031 in England and Wales. Although mean life span has increased, maximum life span has remained essentially unchanged. *Source*: Max Roser, Esteban Ortiz-Ospina, and Hannah Ritchie, "Life Expectancy," Our World in Data, 2013, https://ourworldindata.org/life-expectancy (CC BY-SA), with data from the UK Office for National Statistics.

fall. We have already seen the first wave of this real-time risk monitoring—for instance, in "safe driver" policies. Machine learning will increasingly be able to discern patterns and make predictions from very different data, including proxy data, beyond just underwriting ever more granular health risks revealed by genomic analysis. Policies will reflect your financial risk profile, your driving, your penchant for sports, your level of natural-disaster or crime risk based on where you live.

All of these new data, of course, are highly disruptive in exactly the same way that disruption has ripped through other industries. The large, stable, highly regulated, and bureaucratic ecosystem of insurance carriers, agents, brokers, reinsurance providers, and consultants is shaking as it confronts a wave of startups with quirky names, such as Lemonade, Root, Vouch, Slice, Hippo, and Trov. These insurgents have a quirky label, too—*insurtech*—and they offer mobile apps or artificial intelligence or novel data-driven ways to underwrite policies; a few provide all these things—in insurance jargon, "the full stack." Most will fail or find themselves absorbed by larger carriers. Yet, in aggregate, they are already forcing large insurers to develop far more sophisticated and accurate predictive algorithms using increasingly novel and personalized data sets. The incumbents have already embraced mobile apps and internet interfaces for everything from sales to automated damage analysis, and some are offering almost real-time claims processing. They have also picked up the insurgents' flexibility, particularly in pricing—as the ad says, giving you only what you need without charging you for what you don't.

That's the good news. Increasing data and improving algorithms will lead to more precise and diverse prediction, and potentially lower burdens. Risk may well decline, and prices may as well. But there are negatives, ranging from the moral-hazard consequences of risk reduction to issues of transparency and control and thorny questions about privacy. These themes will recur in the chapters that follow.

Insurance companies, with their forbidding bureaucracies and opaque actuarial underpinnings, are famously unloved. Although flexibility, real-time sales, low prices, and faster claim resolution may make insurance easier and less expensive, they also raise issues that are part and parcel of any ambitious predictive enterprise. Who really understood the actuarial calculations that went into pricing insurance in the first place? As we've seen, for centuries even providers of products such as annuities made fundamental errors in pricing—often, ironically, in favor of annuitants, though hardly in favor of the taxpayers who footed the bills. However, with today's outpouring of data and algorithms, underwriting has grown increasingly precise and personal. But how do buyers of insurance, not to say regulators, know what lies beneath a price? As insurance is ever more deeply woven into everyday digital life, who, including regulators, will really know whether the eternal flow of risk—some calm, some bubbling along, some turbulent—is being accurately or wisely priced? For one thing, it's a flow, not a single calculation. Second, the data are being processed by machine-learning systems, which have great strengths but can resemble black boxes. We may only replace one opacity with another.

Privacy and Moral Hazard

That lack of clarity leads us to a broader issue: privacy. Currently, a significant portion of the population is happy to trade some privacy for a better price or product. For example, we might agree to a medical exam before getting a life insurance policy. But how far does this go, even in terms of health? As medicine learns to truly exploit genomic sequences and other cutting-edge tools to predict probabilities of future health, will consumers accede to insurers' demands to access their most intimate data, particularly if there's a chance their premium could rise after they allow such access? Will consumers resist allowing insurers to get real-time data feeds from their cars that will reveal how fast they drive? Despite the emphasis on individual data, will entire groups—say, those with the possibility of early-onset dementia—find themselves priced out of the insurance market because their future appears probabilistically so grim? And at what point will individuals rebel against the constraints of insurance?

Last, consider the question of moral hazard, which today broadly—sometimes too broadly—refers to the reshaping of behavior, nearly always negatively, because of the reduction of risk. In one of our earliest conversations, Igor called it "Google Maps syndrome." Since the introduction of digital mapping, drivers have eliminated the time they used to spend consulting maps, asking for directions, or just getting lost. Google almost perfectly estimates the time it will take to drive somewhere, so people reasonably leave less and less time for error. The problem here is that by reducing the margin of safety, travelers run other risks that may occur randomly:

an accident or breakdown ahead, a sudden storm, an unanticipated traffic jam, car trouble.

Google Maps syndrome is a variant of moral hazard. Technically, this carelessness or negligence is known as *"morale* hazard," as opposed to the more calculating and deliberate attempt to game insurance known as moral hazard. But as Powers notes, most economists don't make these fine distinctions. The observation that some individuals will change their behavior if they are insured has probably been around since marine insurance first emerged in the third millennium BCE. It wasn't called *moral hazard*, then, which is a term closely associated with the world that first gave us probability and statistics. *Hazard* entered the English language with the Norman Conquest to describe a game of dice; in his dictionary in 1755, Samuel Johnson defined *hazard* as "chance; accident; fortuitous happening."

The use of the word *moral* in this case has little to do with morality or ethics—a matter of some rhetorical confusion. Instead, it emerged from the same set of ideas that resulted in the eighteenth-century mathematical and probability thinkers insisting they were engaged in a "moral science," roughly synonymous with what we now call "social science." Daniel Bernoulli, for instance, referred to "moral expectation" to describe his concept of utilities. As the paper "A History of the Term 'Moral Hazard'" (2012) notes, "Bernoulli's assertion that risks of equal mathematical expectation depending on 'the particular circumstance of the person making the estimate' made a seminal contribution to the theory of risk." This conceptual milestone was critical to the eventual development of a theory of moral hazard that "predicts changes in individual behavior following the purchase of insurance." Moral hazard has proved to be a flexible concept. Because insurance reduces risk, it may also result in a reduction of preventive measures, thus paradoxically nurturing other risks. It's the logic behind the decision not to buy fire alarms if you have insurance. It's the notion that a bailout of banks in a financial crisis will only create incentives for them to act with greater recklessness in the future. It's even been applied to the belief that welfare creates poverty, that workplace regulations spawn accidents, or, as we saw during the COVID-19 pandemic, that too generous unemployment benefits will feed joblessness.

The history of the term—the "genealogy of moral hazard," as one study called it—is an example of how attitudes about prediction and risk have evolved.

The term *moral hazard* emerged in the nineteenth century as the insurance industry rapidly expanded, and it was initially associated with insurers weeding out customers with "bad" character—moral hazards—who were liable to practice bad habits (such as arson or drug use) or engage in fraudulent, loss-making ends. The development of this notion led to several less than quantitative measures to determine a good risk, including physiognomy, the "science" of linking physical traits to moral behavior. Its practical effect was a systematic bias against minorities, foreigners, outsiders, the unconventional, and those broadly viewed for whatever reason as moral risks. This was not just prejudice. The mysteries of prediction and risk, probabilities and chance, had made insurance a suspect business to the public in its early years; insurance scams and failures didn't help. In his article "On the Genealogy of Moral Hazard," the University of Pennsylvania law professor Tom Baker notes that insurers in the Victorian age felt the need to convince the public that insurance was above reproach, "not a form of gambling, a handmaiden to crime, or above all, a presumptuous interference with Divine Providence itself," but rather a scientific product designed for those who deserved it. One way to do so was to embrace the virtues of the day by selling only to the "right people" and thus avoiding the perils of moral hazard.

By the twentieth century, insurance's dubious reputation had been replaced by a conservative, even stuffy image. Bias continued, but it was less overt and often justified by blaming actuaries and their mysterious calculations. By the late 1960s, the meaning of the term *moral hazard* had shifted again. In 1963, the Stanford economist Kenneth Arrow published a seminal paper, "Uncertainty and the Welfare Economics of Medical Care," that analyzed the economics of the American health-care system, with a particular focus on the insurance market for medical care. In effect, Arrow brought medical markets, including insurance, into economic models that featured competing, maximizing agents; this was new. Arrow argued that medical insurance was particularly plagued by moral hazard and as a result markets had failed to provide insurance, either for significant portions of the population or for certain kinds of medical problems. Health and medical services featured a great deal of uncertainty; they were, as a result, difficult to predict. One manifestation of that uncertainty is a paradox, Arrow wrote: "Widespread medical insurance increases the demand for medical care." Given uncertainty and a resulting unpredictability, there was little incentive for consumers or their physicians to reduce costs. Arrow recognized that while individuals cannot control many illnesses, they can pick their doctors, who have no reason not to provide a full range of services. Insurers, in turn, respond to what they perceive to be a moral hazard by restricting coverage or by raising prices.

Given all this, Arrow concluded that the "welfare case for insurance of all sorts is overwhelming," and that the government should provide insurance "where the market, for whatever reason, has failed to emerge." Arrow's paper was published two years before Congress passed Medicare.

In 1968, Mark Pauly, a health economist then at Northwestern University, replied to Arrow's paper, providing a somewhat different view of incentives and moral hazard, as well as a different conclusion. He argued that moral hazard was less a question of morality than a rational response to subsidies and the clash between the collective need for lower medical costs and an individual's desire for more care—that is, it was an economic problem. He compared the situation to game theory's prisoner's dilemma, in which two rational prisoners, each seeking the best deal, inadvertently betray each other. And he argued that "some uncertain medical expenses will not and should not be insured in an optimal [economic] situation. No single insurance policy is 'best' or 'most efficient' for a whole population of diverse tastes." In short, Pauly's comment undermined the argument for universal coverage. Arrow responded to Pauly, in turn, arguing that medical market incentives alone did not determine consumer choice, and that for the economy to properly function there had to be a degree of trust and professional obligations between the principal and the health-care agents.

Pauly disagreed, and on that issue he proved prescient. More and more, health care became a business. Over the next 50 years, moral hazard became a matter of external incentives rather than internal moral considerations. As costs rose, economists and policy makers increasingly viewed health care as less about trust and service—what in health-care economics has been called "the Marcus Welby medical economy"—and more about efficiency, market forces, and incentives to reduce use, such as deductibles and copays. Arrow himself never fully lost his belief in a less competitive system. Today, moral hazard has little to do with moral probity and more to do with the incentives insurance creates and the behavior it engenders. Moreover, economics increasingly has embraced the term to include the consequences of a range of principal-and-agent issues, such as social welfare or regulation, arguing in effect that welfare creates poverty or that regulation ensures more, not less, abusive behavior.

One result of this debate over what became known as welfare economics, however, was to reveal how complex consumers' response to insurance actually is, particularly as medicine and insurance become more personalized and predictive. That complexity is just the start. Now that we can measure more than ever before, we can predict better, and the entire system is poised for change.

"Everything is different now. We can solve almost any crime, anywhere."

That's the audacious claim of the geneticist and forensics expert David Mittelman, who built the first private forensics genomics laboratory, Othram, to bring the power of modern genome sequencing to forensic evidence. Located just north of Houston, the company has one of the world's cleanest "clean rooms," necessary for scraping every fragment of DNA from any given crime scene sample and ensuring almost no contamination, with the goal of revealing myriad hidden clues about who did what and when. Othram's motto matches its ambition: "Justice through genomics." For a company that launched in 2018, it has an extraordinary track record, having already helped solve hundreds of cold cases, including some seemingly impossible cases from the 1950s.

Mittelman, Othram's CEO and cofounder, resembles a bodybuilder fresh from a Mr. Universe contest yet is gentle, curious, and friendly. He wants to improve the way crimes are solved. His passion for forensic science stems from his belief that the world today has a substantially different predictive capacity than ever before that can be used to reveal unprecedented molecular narratives of individuals who leave their DNA behind. "Our DNA markers tell a story, and we are just now finally starting to listen," he says.

Othram's team notes that established forensic testing frameworks, such as the FBI's CODIS (Combined DNA Index System), are inadequate for solving many older crimes and were never designed to tackle the challenge of unidentified human remains, which primarily belong to victims and not perpetrators. CODIS DNA profiles, known as "DNA types," are stored in a database after FBI technicians genotype parts of forensic DNA. These genotyping efforts focus on short tandem repeats (STRs) in the human genome, simple sequences that repeat the same sequence of 100 to 300 bases. To perform an analysis, DNA is extracted from a sample; the region containing each STR is amplified by a polymerase chain reaction, which massively multiplies the sample. The set of repeats is then examined for its size and pattern, thus creating a person's STR type. The STR type is an "allele," which is an individual's version of a genetic sequence or gene; normally everyone gets one allele from each parent. As such, an individual with a specific set of STR repeats or alleles will share 50 percent of them with siblings, 50 percent with each parent, 25 percent with their grandparents, and 12.5 percent with their cousins, and so on.

Beyond similarities within families, a specific STR profile *should be* relatively unique. The CODIS forensics system today is based on 20 STRs, which should produce a relatively unique combination for each person. This assumes that all 20 STRs are scattered widely across the genome and don't move together, which is usually true, and that the population randomly mates, which is somewhat true. The FBI-determined STR allele frequencies generate a statistical prediction that the chance of two unrelated Caucasians having identical STR profiles, or DNA types, is approximately one in 575 trillion.

Though such a chance sounds quite rare, it's important to note that this number is for matches of *pairs of people*—and there are a lot of possible pairs of people in this world. Matching N people to the number of possible pairs would involve $(N^*(N - 1))/2$ possible combinations. For example, if one were to look at 100 million Caucasians across the world, there would be 4,999 trillion possible pairs and 8.69 pairs that match by chance alone. For most crimes, this might be sufficiently precise, but it remains an imperfect system for cases that involve serious crimes and long prison sentences and is wholly unsuitable for identifying victims of crimes.

Othram uses more markers than CODIS does, including a method of so-called shotgun sequencing and enrichment, like the swabbing and sequencing methods of the New York City subways performed by Chris's lab at Cornell, which looked at hundreds of millions of DNA fragments, not just 20, to build a profile. This method changes the profile of each person from a yes/no match (if that person's profile is in the FBI database) to a much more dynamic and wide-ranging profile of anyone who leaves his or her DNA behind—that is, potentially everyone.

Ubiquitous Genomic Data

DNA testing and other molecular forensics technologies have transformed traditional forensics, enabling predictions about guilt and innocence based not only on DNA left at the scene of the crime but also on the analysis of DNA that could have been found anywhere. It need not be the result of a crime. We all leave DNA on every surface we touch; all an employer needs to do is shake your hand or wait for you to leave a cup behind, grab a sample, and draw predictive conclusions based on your DNA. This DNA may reveal not only your disease risk, traits, and facial features but also your epigenetic age, based on technologies for DNA methylation analysis. Each surface offers clues to where you've been and what you've been doing.

Every investigation is, in the end, an attempt at prediction, and now we can do it with methods that were impossible only a few years ago. Recent advances in the utilization of genetic technologies have begun to change the very nature of forensics, albeit at a cost: the narrowing of reasonable expectations of privacy. Today, investigators can tap vast new pools of data, from networks of security cameras to millions of devices in the IoT, from social media and smartphone videos and internet searches to telltale genomic identifiers. As more people take genetic tests from companies such as 23andMe, Onegevity, MyHeritage, and Ancestry.com, more people can be matched with their relatives. Making sense of all this are a number of AI tools that employ algorithms used in computer vision and graphics, data visualization, and machine-learning programs to discern patterns in images, documents, addresses, electronic records, and biological evidence— the kind of genomic evidence that led to the capture of the Golden State Killer a few years ago after a decades-long hunt.

However, the toolbox of forensics and genomics can also be used as a weapon, making identity tracing and genetic discrimination easy, and opening up a particularly thorny set of issues around aggressive predictive tools.

Such concerns have been building for a while. In the movie *Gattaca* (1997), widespread genetic profiling, biometric identification, and institutionalized eugenics have led to the normalization of discrimination based on genetic makeup. *Gattaca*'s fictional society is grounded on the notion that an individual's fate is not only determined at birth but also reinforced by continual, ubiquitous DNA sequencing and surveillance made possible by rapid genetic profiling at every stage of life. This profiling includes routine testing of embryos prior to in vitro fertilization (known as preim-plantation genetic diagnosis), rapid genetic testing of potential mates (to assess long-term "dating potential"), and job stratification and placement.

In the late 1990s, such routine and pervasive genetic profiling was far too technologically challenging and costly to be realistic. As such, *Gattaca* was viewed largely as a cautionary tale about genetic discrimination and the dangers of anchoring a multitude of pivotal life decisions on DNA alone. Yet the film's depiction of issues pertaining to privacy, genetic discrimina-tion, and genetic engineering can be seen as prophetic because the methods and tools of diagnostics and surveillance it features are now a reality.

Around the time *Gattaca* was hitting theaters, the debate over genetic pri-vacy began to simmer, fueled by new genomic tools and the development of genetic testing. The first headline case involved professional basketball.

Private and Public

In 1987, the Boston Celtics drafted Reggie Lewis, who in 1992 became a National Basketball Association all-star. But in April 1993, Lewis suddenly collapsed during a game. He was diagnosed with rare genetic mutations that conferred a risk of hypertrophic cardiomyopathy, caused by an abnor-mally thick heart muscle that can make it harder to pump blood. Carriers of the mutations in any one of 27 specific genes have a higher chance of death from arrhythmia, valve problems, and sudden heart failure. At the time, however, there were few easy ways to do the kind of genetic testing necessary to diagnose the condition. Later in 1993, Lewis again collapsed during an unofficial practice. This time he died. The shock reverberated throughout the NBA.

After the Michael Jordan era, the Chicago Bulls needed some new play-ers, and they signed a promising young player named Eddy Curry. Late in the 2004–2005 season, Curry was hospitalized with an irregular heartbeat. He missed the playoffs, but doctors then cleared him for playing. Because of Curry's heart issues, the Bulls management asked him to take a genetic test to prove he did not carry hypertrophic cardiomyopathy mutations. Curry resisted, asserting that such testing was an invasion of his privacy and an unethical breach of his medical information. The Bulls management insisted that it did not have an ulterior motive beyond concern for Curry, and they offered him an annuity of $400,000 per year for 50 years if he "failed" the genetic test and carried some mutations. But, he refused again, and his medical team insisted that his cardiac issues had been resolved. The Bulls then traded him to the New York Knicks, and he was traded several more times over the subsequent years. Curry's career sputtered, but he never had a recurrence of the irregular heartbeat.

Around the same time, another case of alleged genetic discrimination emerged at the Burlington Northern and Santa Fe Railway (BNSF), one of the largest US rail companies.

As at most businesses, railway employees were increasingly working on computer keyboards, and a growing number had begun to request medical support for carpal tunnel syndrome, a common ailment among office workers. The carpal tunnel is a narrow passageway surrounded by bones and ligaments on the palm side of the hand; if the median nerve is compressed—by excessive typing, say—people can experience numbness, tingling, and weakness in the hand and arm. The condition can make it impossible to work on a computer. BNSF employees filed some 125 carpal tunnel claims in 2000–2001, and more were expected in 2002.

As with hypertrophic cardiomyopathy, some manifestations of carpal tunnel syndrome have a genetic component. BNSF's management saw genetic testing as a way to identify potential "problem employees" and to avoid responsibility, both occupationally and medically, for its employees. If the company could "prove" that some employees were genetically at risk and likely to have the syndrome, it might not have to pay workers' compensation. Management eventually asked 36 employees for blood samples, with no informed consent. The workers' union then filed charges with the Equal Employment Opportunity Commission (EEOC), which filed suit in a federal district court in Iowa. The EEOC argued that the case represented violations of the Americans with Disabilities Act (ADA) of 1990, which bars discrimination based on disability. However, carrying a genetic risk is not technically a disability—rather, it's a probability of a range of ability—so the EEOC's argument was a risky legal move.

But it worked. The suit was settled through mediation in 2002, with the railroad agreeing to pay a total of $2.2 million to the 36 employees for violating the ADA by genetically testing or seeking to test them without their knowledge or consent. The settlement stated that BNSF would not directly or indirectly require employees to submit blood for genetic tests; analyze any blood previously obtained; evaluate, analyze, or consider any gene test previously performed on any of its employees; or retaliate or threaten to take any adverse action against any person who opposed the genetic testing or participated in the EEOC's proceedings.

By then, advocacy groups and lawyers had begun calling for an updating of the laws around discrimination, genetics, and privacy. This was the genesis of the Genetic Information Nondiscrimination Act (GINA), passed in 2008. Title I of the law prohibits issuers of health insurance from discriminating on the basis of genetic information for eligibility, coverage, underwriting, or premium-setting decisions. Title II prevents employers from using genetic information in employment decisions, such as hiring, firing, promotions, pay, and job assignments. Lastly, Title III prohibits employers or other covered entities, such as unions, from requiring or requesting genetic information and/or genetic tests as a condition of employment.

The Devious Defecator

The law's first big test came in 2012, when an Atlanta grocery distributor, Atlas Logistics Group Retail Services, could not figure out who was leaving feces on the floor of a warehouse. Atlas was looking to track down what appeared to be a prankster employee. First, the company tried to use time sheets and video to find the culprit. Failing at that, it asked two employees for their DNA; they felt they had to comply to requests for cheek swabs for testing, or they would be fired. Both DNA tests came back negative, so the source of the defecations remained unknown.

The men filed a lawsuit against Atlas Logistics—a case that was poetically dubbed the "mystery of the devious defecator" by US District Court judge Amy Totenberg. Atlas argued it was simply trying to keep a safe workplace, while plaintiffs argued they had been singled out, accused, and illegally tested, losing time, money, and reputation. The case went on for two years. In 2015, a Georgia jury awarded $2.25 million in damages to the two men. Similar court cases have kept appearing; since 2010, the EEOC has filed 200 to 300 cases per year under GINA.

The law is hardly perfect. It does not apply to employers with fewer than 15 employees because legislators thought compliance would be too onerous for small companies. Also, the law's protections do not extend to the US military, the Tricare military health system, the Indian Health Service, the

Veterans Health Administration, or the Federal Employees Health Benefits Program.

More importantly, by the time GINA passed, the technology, particularly next-generation sequencing (NGS), was already racing past it. As we saw in chapter 3, DNA sequencing is now inexpensive enough that it can be employed at every crime scene, in every bedroom, toilet, or turnstile. Access to this type of biological material and genetic information can identify location and trace movements with a DNA sequencer the size of a suitcase—a process dubbed "ubiquitous sequencing."

But is it legal to sequence DNA found in a public space? How do we draw the line between public and private? The US legal system's notion of privacy is based on "what a person *knowingly* exposes to the public." The US Supreme Court stipulates that whenever people are aware that they are exposing something to the public, even in their home or office, that something is "not subject to protections from the Fourth Amendment." The Fourth Amendment protects Americans from "unreasonable searches and seizures" in their "persons, houses, papers, and effects."

Some legal precedents set limits on privacy. Take, for instance, the case of Danny Kyllo, who was arrested for growing large amounts of marijuana at home after the police used thermographic cameras to observe the heat emanating from his house. Even though the police got a search warrant to use the special camera, Kyllo challenged this search because most people cannot "see" within the infrared range of light. Thus, Kyllo argued, the police had violated his right to privacy in the infrared range of the electromagnetic spectrum, and the police's use of the cameras violated his Fourth Amendment rights. The Supreme Court ruled in favor of Kyllo in 2001, arguing that the Constitution protected his right to privacy, while maintaining the requirement of a search warrant to monitor individuals.

Despite Kyllo's victory, such cases illustrate the fact that expectations of privacy are both time and context specific. If the Kyllo case were argued today, the result would not necessarily be the same. Reasonable expectations of privacy constantly shift. Once a technology becomes ubiquitous and inexpensive, it becomes *reasonable* to assume someone can use that technology to monitor someone else's activity.

In 2016, an executive order issued by President Barack Obama established the Federal Privacy Council, which aimed to update and coordinate privacy measures across the government. Obama's order was prompted by several incidents, including the rise of warrantless wiretapping and highly accurate facial-recognition software used in an FBI database that contained photos of about one-third of all Americans. In the context of genetics, these privacy concerns can have an even graver impact because they concern both individuals and entire families. Although GINA outlawed genetic discrimination for health insurance and employment, a person's genetic profile (or that of a relative) can still be utilized for *legal* genetic discrimination in denying or adjusting life, long-term care, and disability insurance. Some genetic discrimination is legal if it is based on DNA people have left behind. Thus, the key question recurs: When people leave DNA behind wherever they go, what exactly are they knowingly exposing?

Molecular Signatures

What can we find in a pile of DNA left behind at a crime scene? A great deal (figure 7.1).

With NGS devices getting smaller and more portable, sequencing can be done anywhere and anytime, such as during an Ebola disease outbreak, or even on the space station. NGS has enabled us not only to sequence DNA but to go beyond it; this makes the Age of Prediction a postgenomic era. We are using these technologies to investigate all aspects of the central dogma of molecular biology, including gene expression, DNA structure, and regulation, and to explore areas of human biology such as the dynamic epigenome, which has its own epigenetic age, and the landscape of RNA modifications and regulation known as the *epitranscriptome*.

Additional individual-identifiable data have also come from the *microbiome*, or the vast diversity of microorganisms that are in, on, and around us, including bacteria, viruses, protozoa, and fungi residing within or on the body, and from the *metagenome*, the genome across all species. Recent estimates suggest that there are likely as many bacterial cells alive in our bodies as there are human cells—sometimes more. NGS has been used to discern where these organisms are located in the human body, creating a kind of molecular cartography. Investigating the microbiome has also enabled the discovery of new drugs, and recent work in metagenomics has examined life across kingdoms, including urban genetic maps of cities around the world (the MetaSUB project), a comprehensive microbiome map of Earth's

Suspect's Molecular Profile

Test	Result	Accuracy	Analyzed Molecules			
Physical Characteristics						
Ancestry	Wisconsin	99%	M		G	
Age (biological)	40	90%	M	mG	G	
Age (molecular)	27	99%	M	mG	G*	T
Eye color	Blue	99%			G	
Height	6'	99%			G	
Weight	82kg	99%	M	mG	G	T
Mood	Happy	75%	M	mG	G	T*
Last Meal						
Food	Bananas	99%	M	mG	G	T
Beverage	Whiskey	99%	M	mG		T
Diseases						
Inherited	Huntington's	99%			G	
Chronic	AIDS	99%	M	mG	G	T
Acute	COVID-19	99%	M	mG		T

Samples Collected
Hair follicles
Counter and floor
Saliva (5uL)
Rim of plastic cup

Metabolome, Metagenome, Genome, Transcriptome, *(Epi-)

Figure 7.1

Cross-kingdom methods of forensics. Categories of multiomic and multikingdom measurements can create both a forensic profile and a social profile for a person based on that person's metabolome (M), metagenome (mG), genome (G), epigenome (*G), transcriptome (T), and epitranscriptome (*T). Categories for types of inferred information are detailed by the trait/activity and the revelatory information.

cities, as well as an analysis of extremophiles, microorganisms that live in extreme conditions.

Given all of this, what can a fragment of DNA left on a surface tell us? Published data indicate that at least 11 phenotypes can be tracked with at least 16 methods, spanning many areas of molecular profiling: ancestry, facial features, age, identity, cell type, environmental history, geospatial localization, obesity, presence of a disease or infection, circadian rhythm, and pregnancy. These phenotypes can be detected across five categories of molecule: human DNA, human RNA, epigenetics, epitranscriptomes, and metagenomes. Each of these categories produces its own identifying data that can be used both medically and forensically.

The first molecular target is the best known: human DNA. Researchers have been able to use DNA to construct the map of Europe from ancestry-informative markers (AIMs), while placing genetic information in a "data continent" that resembles the actual geographical one. These maps can help pinpoint a birth city within a few hundred kilometers. With recent tools such as DNA.Land (by Yaniv Erlich and Joe Pickrell) and genomic and metagenomic GPS built from MetaSUB data, researchers can subclassify ancestry and location anywhere in the world with even greater accuracy. With these AIMs and other genetic markers, we can predict likely hair type, eye color, skin darkness, and other physical features, such as earlobe type, dimples, mouth shape, or a widow's peak. Human Longevity, a genomics startup launched by Craig Venter and Peter Diamandis, has shown that we can also predict someone's voice, height, and facial features with high degrees of precision. Other data have shown the ability to detect gray hair, beard thickness, even the shape of a head of hair.

Still other techniques offer the ability to predict age based on DNA. Most people have a set of 46 chromosomes, with 23 each from their mother and father. But this number changes over time, with cells losing chromosomes as a person ages. Every time cells replicate, errors can occur, and the likelihood of such errors gradually increases over time. Men's cells slowly lose their Y chromosomes, while both women and men lose their X chromosomes. As we age, the telomeres at the ends of our chromosomes steadily shrink, and new mutations emerge in our bone marrow from what's known as clonal hematopoiesis, producing a population of mutant blood cells that may anticipate cancer or other diseases of aging. Knowledge of these various dynamics provides a way to compute a person's age and risk of long-term disease.

Moreover, identity can be traced. STRs from the Y chromosome can reveal the surname of a person who contributed a supposedly anonymous data set to a public archive. These markers can be used to find a relative, who if he or she has ever been arrested is likely to show up in the CODIS database, just as a familial DNA search led to the Golden State Killer. In general, you should not commit crimes. But, if you have a relative who has been arrested, it's a *particularly* bad idea for you to commit a crime, since part of your genetic code is already in the database.

Beyond human DNA, there is RNA, the active form of DNA. Here, too, the transcription level—that is, the amount of RNA—is dependent on gene regulation, which usually occurs at the level of DNA and is a reflection of ancestry. These "control boxes" are called "quantitative trait loci" (QTLs); when they are linked to gene expression, they are called eQTLs. Amazingly, an individual's levels of gene expression can reveal what genetic variants that person is carrying and by extension that person's ancestry.

RNA also tells a story on a cell-by-cell basis. Each cell type has a specific and unique expression program, so that measuring RNA can tell you the cells of origin. This includes the ability to distinguish between different areas of your skin and can reveal what part of your skin touched a surface.

The third forensics category is epigenetics, which are changes in gene expression that don't involve changes in the DNA sequence. In an embryo, all DNA methylation markers are reset to give that first cell the full capacity to differentiate into any cell in the body. But this lineage specification of cells is not permanent; certain areas of your genome slowly lose methylation, and some slowly gain it. In fact, there are 353 specific sites in your body that can almost perfectly predict your age, and more have been discovered by research fellows at Weill Cornell on loan from Igor's firm, WorldQuant. Chris's lab at Weill Cornell used these methods to solve part of the case of Steven Avery, the subject of the television documentary series *Making a Murderer* (2015). Avery accused the police of planting old blood at the scene of a crime, but epigenetic methods from Chris's lab showed it was actually fresh blood, and the legal teams had to change their arguments.

Another aspect of epigenetics is how DNA is wrapped or packaged inside cells. These chromatin states have been shown to be heritable and potentially distinct for different ethnic groups. Analyzing those different states also works in primates, whales, and other species, although these epigenetic differences are also a reflection of the proportion of cell types in a tissue as it ages. As a result, these epigenetic marks can be used to discern family and genetic background.

If RNA doesn't tell you enough about the type of cell present, you can use DNA's epigenetic signatures as clues. Because small epigenetic marks define the function of genes in cells and thus cell types, measuring epigenetic marks can provide clues as to which cells are present. This is important for cancer research on metastasis—the spread of cancer cells from their organs of origin—because even a rogue cancer cell that is on the move still carries its epigenetic history and holds clues about the organ from which the original tumor developed. These differences in cell types may also explain why some tumors develop specific kinds of mutations.

The fourth class of identifying molecules comes from nonhuman parts of DNA and RNA—the microbiome and the metagenome. Sampling a surface can reveal a shift in an ecosystem that has persisted over years. The MetaSUB project saw this after the South Ferry subway station in New York City was flooded during Hurricane Sandy in 2012; more than a year after the storm, Chris, Evan Afshin, and colleagues found that the walls of the station still contained evidence of marine and water-based microbial life, as a kind of molecular echo.

Forensic geology is a field that uses the nonrandom distribution of geological elements, chemicals, and mineral deposits to ascertain the geospatial provenance of samples. Toward the end of World War II, for instance, forensic geology was used to prove that the Japanese had launched a series of secret balloon bombs on some western American states. The investigators tracked down sand used in the ballast bags to the west coast of Honshu, Japan, which had beaches of that exact sand; the technology has since then been used in crime scene forensics. Now *metagenomic* forensic methods can potentially tell where on your body a hair came from, identify your gender based on sex-enriched bacteria, and even say whether the soil on your shoe came from a park or a median near the park.

One of the targets of metagenomic analysis is the microbiome—specifically, the genomes of anything microscopic. Microbiome profiles of obese and skinny people differ in their fungal and bacterial markers. These molecular differences reveal important information on how individuals digest and process food that can be linked to a person's identity. Also, each human gut appears to have a specific metagenomic genotype. The gut metagenome is thus similar to human DNA in the sense that it can be used to determine location; this includes matching your microbiome identity to your personal computer even if you haven't touched it in several weeks. If you set up a DNA sequencer in every toilet, you could readily track people's movements due to their "gut print" and identify the part of the body that touched various surfaces based on a molecular cartography of the body.

Collections of microorganisms sometimes include pathogenic bacteria responsible for the spread of infections. These bacteria can be used to predict people's sickness phenotypes (observable sickness characteristics or traits), and they can be used in "smart" sewage systems that can provide early warnings of infectious-disease outbreaks. Several towns and colleges have used such systems to try to control the spread of COVID-19. In 2015, MIT began Underworlds, a sewage census of cities around the world to map urban epidemiology. The fact is that different microorganisms within the human microbiome tend to gather in different areas of a city, and macroorganisms have a phylogenetic structure across a city.

Even the placenta, once thought to be a sterile organ, can sometimes harbor bacterial species such as *Streptococcus agalactiae* (Strep B), which is normally quite benign in the gut but can cause a nasty infection in infants, the elderly, and the immune compromised. If sequencers were installed on every toilet in a geographical area, we could determine the respective locations of pregnant women. Some creative attempts have been tried without plumbing. In a classic example of proxy data, in 2012 the retailer Target developed an algorithm that predicted which women shoppers were pregnant based on their purchases and demographic details. The Target computer focused on 25 products associated with nascent pregnancies, which when analyzed together produced a pregnancy-prediction score. The retailer sent flyers to customers with a high score, in one case inadvertently revealing a woman's pregnancy before she told her family.

The fifth type of molecule with forensic potential is the epitranscriptome, which describes the biochemical modifications of RNA. Despite its relatively recent discovery in 2012 by Chris, Samie Jaffrey, and colleagues, the epitranscriptome is likely to be as important to RNA regulation as epigenetics is to DNA regulation. In fact, it's possible to use the presence of specific epitranscriptomic markers, such as methyl-6-adenosine, to determine if someone slept in a bed or just left cells on the sheets while awake due to the cells' tight coupling with circadian rhythm. Assuming RNA is preserved on a surface, modifications to RNA can indicate if a person's body is responding to an HIV infection as well as the type and severity of the infection.

Of note, many of the other advances in genomics and biomedicine have recently come from improved cameras and imaging. These tools have enabled us to peer inside single cells with unparalleled resolution, and the low cost and high resolution of cameras have enabled their ubiquity and applications in forensics. By using drones that maintain a geosynchronous location above a city, you can make a "time machine" of movement by taking a picture every minute of every day. The drones are made by the company Persistent Surveillance Systems and were originally used for tracking people planting roadside bombs in Iraq. They have also tracked and led to the arrest of alleged criminals in Dayton, Ohio, and Juárez, Mexico, and dramatically changed the nature of warfare, the subject of chapter 8.

This kind of surveillance has serious implications for expectations of privacy. Movements can be tracked in criminal investigations or, for less uplifting ends, in marital disputes or employment cases. Combined with data-mining methods, such as Target's pregnancy-detection algorithm, these types of surveillance tools are eroding traditional notions of privacy.

Not surprisingly, these encroachments on privacy have also produced new ways to obfuscate; some methods of obfuscation have even been patented. To mask your presence, synthetic oligonucleotides can be sprayed behind any chair or area you occupy. However, this "genetic camouflage" works only if it perfectly matches your genome; if it doesn't, it can be used as evidence against you. The synthetic DNA would need to match the real DNA, with identical DNA methylation and epigenetic groups, as well as microbiome and RNA profiles.

Yet the prospect of an ability to create such a precise molecular match leads to problems. Technology for accurate and specific matches at the genetic, epigenetic, RNA, epitranscriptome, and microbial levels makes it very easy to frame anyone for a crime as long as you have full access to that person's biomolecules. Moreover, with precise methods for transdifferentiation and redifferentiation of cells, it might be possible to convert skin cells left behind into red blood cells to suggest that blood had been spilled. With CRISPR (clustered regularly interspaced short palindromic repeats) genomic-editing systems and epigenetic-modification mechanisms, precise and potentially malicious genetic manipulations are possible.

This is the new world of forensics. These powerful new tools and scant privacy protections may turn out to be the greatest threat to both the new forensics and to precision medicine, epidemiology, and real-time insurance because they create impediments—legislative, legal, personal—to the free creation and sharing of data. Ironically, our unprecedented ability to sequence entire genomes has simultaneously spawned personal concerns about what those data may reveal to law enforcement, insurers, and employers, concerns that may significantly affect large-scale attempts to use those data to reduce crime and track everything from pandemics to migrant flows to cancers. This kind of data is a little like free trade, which for all its ability to create wealth globally, has resulted in balkanization, trade wars, tariffs, economic decoupling, and belligerent nationalism. The resistance to sharing data is just one potential paradoxical consequence of these powerful tools of prediction. Beyond that, the new tools create even more dystopian possibilities in autocratic surveillance states, which are called police states for a reason.

GINA 2.0

Without question, many of these predictive tools are probabilistic, and genotyping methods and interpretation can never be 100 percent accurate. But given the number of possible tests across different data types, metrics, and kingdoms of life, a new genomic forensic landscape has emerged that turns previous notions of privacy to rubble. The speed and easy availability of these tools, algorithms, and many-layered mechanisms for detection and reprogramming of identity from genomics and postgenomic data will create challenges for judicious enforcement because you can quite literally change the markers of your identity. They also raise questions about who has access to metadata, genetic material, and the potentially revelatory information contained in microscopic particles left on every surface. Such information and metadata pose a significant risk of discrimination. For example, while GINA covers employment and health insurance discrimination, it does not prevent life insurance underwriters from changing premiums based on genetic markers. Moreover, any information about family members could also *legally* be used as a basis for altered eligibility, coverage, or premiums on life, disability, or long-term-care insurance. By extension, any of these forensic mechanisms may be used to change, deny, or alter coverage for a person or that person's relatives.

Also, GINA was originally designed to protect human DNA and genetic information, but the statute does not necessarily apply to the microbiome or the metagenome. All of the personal identifying information based on microbial signatures can be used for the precise activity that GINA was trying to avoid: health insurance and employment discrimination. Given this loophole, an updated GINA should account for these nonhuman markers as well as for the battery of molecular signatures described in this chapter. Indeed, a more inclusive bill might specify that any "personally identifying molecular signature" be exempt from use in insurance and employment decisions and options for individuals or their relatives.

Along with the current set of existing human and genetic rights, a new GINA could also legally establish certain genetic rights afforded to each person. These rights would include the right to sequence your own DNA, have your doctor sequence your DNA, modify your DNA, amplify your genes, sequence the DNA of things you encounter in the world, interpret your DNA, and leave your cells behind without risk of discrimination. Because there are now so many methods to identify a person, the inability to remain anonymous may be a foregone conclusion; we can only try to ensure that there is no illegal or unethical discrimination.

Because being born is the most universal "preexisting condition," one could argue, using what the political philosopher John Rawls called the "veil of ignorance"—a device for ethical reasoning that involves not knowing the benefit or outcome of a set of conditions—that it is unethical to penalize people for genetics they could not prevent or control; no one can choose the body into which one is born. Yet actuarial calculations for insurers depend on *some* ability to hedge or pool risk across large numbers of people. Thus, absent a system of universal coverage for all individuals for all types of insurance, the question of appropriate utilization of these insurance mechanisms will likely lean toward more expensive and punitive measures for congenital conditions.

However, the greatest reason to resist discrimination based on genetic markers is their fallibility, which is driven by biological complexity generally and by DNA's pleiotropic nature more specifically, where one gene may have multiple functions. The statistical probability of the emergence of a phenotype is *not* the same as the actual manifestation of that trait. The transition from potential trait to actual trait should be acted upon only when it occurs and perhaps only if the "owner" agrees. As things stand now, as soon as you leave a room, you give up all your rights to your library of personal molecular information. And because it is unlikely that hermetically sealed space suits will come into fashion, the goal should be to ensure that discrimination and malicious use of the "leftover" biomaterials that you shed everywhere be banned. However, removing loopholes of currently legal, if unethical, discrimination and mitigating the risk of future discrimination require changes in legislation, in people's expectations of privacy, and in the use of DNA from all species.

So much of the burgeoning digital architecture—the IoT, social media, mobile telephony—has become both a tool for policing and security and a vector for criminals. The algorithm itself has become a tool for good and evil, just like other powerful technologies, from fire to atomic energy to gene editing. These technologies represent a potential evisceration of traditional views of and expectations for privacy. Balanced against this negative cost is the law enforcement benefit. Criminals could be caught more quickly, there might be fewer wrongful prosecutions, and personal responsibility for most public actions would be anchored more securely in data. In the past, people have traded privacy for even the illusion of security. We predict many people will accept a loss of privacy for the actual reduction of risk as long as someone is watching how these algorithms and data are being used. But, even if this prediction is only partly true, this coming world will be dramatically different.

8   The Smart Killing Machine

Chula Vista is a California municipality of 52 square miles and a population of 270,000 tucked between San Diego and the Mexican border. To the west, the Pacific Ocean sparkles; to the south is Tijuana. Over the past few years, the Chula Vista Police Department has used drones built by a San Diego company to support local law enforcement, particularly in emergency situations. Relatively inexpensive drones are regularly dispatched to the sites of fires, car crashes, and crimes. This practice makes good sense in a pandemic, when reducing exposure to the virus is a constant concern. But the drones—far smaller than the Reaper or Predator drones the US military uses for surveillance or missile strikes—raise the same questions of privacy as the persistent surveillance systems we discussed in the previous chapter on forensics. And in one way, Chula Vista has gone where the US military has not: powered by AI software, the police drones have the ability to independently track suspects.

Policing and military organizations share many similar features; they are cultures shaped by the need to achieve security through (if necessary) the legal use of deadly weaponry. Part of the current debate over policing involves police departments' use of military-grade hardware, from armored carriers to assault rifles. Since the dawn of history, there has never been a machine capable of autonomously making life-or-death decisions. The decision to take or save a life has occurred within the human chain of military command. Though their use may be controversial, military drones are operated by human handlers in the United States; unlike self-driving cars, they are deliberately *not* autonomous. And the Chula Vista drones carry cameras, not weaponry.

Nonetheless, the potential to build autonomous missiles, automated-firing technologies, self-guided defense systems, robots that can hunt down improvised explosive devices, and autonomous swarms of air- or land-based robots has arrived. All of these advances, now being explored by militaries around the globe, use some form of AI, which operates by processing constant streams of data to make rapid decisions—that is, predictions. Just as a self-driving car must identify pedestrians crossing an intersection ahead and calculate the best strategy to avoid them, so must AI for military use, processing data about enemy intentions and capabilities, calculate the best strategy to respond to them, to knock them out, and, ideally, to win. The question at this point is not if we engage with such technology, but rather how.

Precision Warfare

Predictions are a raw power with an innate capacity to create as well as to destroy. Perhaps nowhere is that range of divergent possibilities so stark and so wide as in warfare. Precision warfare spawns the comforting belief that armed conflict can proceed with a minimum of civilian and military casualties. In fact, it might be possible for there to be no human casualties at all, with countries waging entirely virtual and mechanical battles and humans controlling the hardware but putting no human lives at risk. Such a future world is not yet here and may never be. Reducing casualties to zero may be a deeply rational and human instinct, but war is not rational or humane; it's about anger, revenge, and terror. A video game war may be expensive to make, but it's relatively antiseptic. It lacks deterrence and coercion; you can easily walk away from it. Real war, as we've rediscovered in Ukraine, is about punishment and power. The same technologies that bring us the seductive possibilities of more precise warfare also bring us autonomous warfare with its unsettling questions about ethics and control.

Like medicine, conventional warfare has grown increasingly precise and, in a sense, personal. Precision-guided missiles and bombs can exact devastation with controls as simple as those of a video game and from thousands of miles away. Telemetry sensors and guidance systems for intercontinental ballistic missiles (ICBMs) can easily target distant cities. Cyberwarfare can disrupt a city or a business or even a single phone with smart weapons.

But this begs the question, what is a "smart" weapon? It's a munition that (in theory) can hit its intended target and nothing else. That's not easy, and there have been many errors, much as a fully autonomous driving system remains a work in progress. Many weapons may be sophisticated but not very smart; they may be automated, like machine guns, but not at all autonomous; they have to be controlled by humans. Thermal- and image-sensing technologies can guide an explosive to a precise target, but they cannot ensure that collateral damage will not occur or, for that matter, that the underlying intelligence was correct. Sometimes, collateral damage is the point. For example, cluster bomb units (CBUs) have been used since World War II to maximize the explosive weight of a single dropped bomb. These bombs are used in strategic attacks that are distinct from carpet- or saturation-bombing strategies, which attempt to destroy as much of a given area as possible. The killing is random, not precise. It's a weapon designed to create terror.

However, recent CBUs have grown more sophisticated and technically more precise. For example, the CBU-97 Sensor Fuzed Weapon, a Textron Systems bomb developed in the late 1990s for the US Air Force, is a 1,000-pound free-fall CBU married to a wind-corrected munitions-dispenser guidance tail kit for targeting. As the bomb falls, its skin pops open, releasing ten submunitions. Each submunition contains a small parachute that slows its descent. If you were to look up, ten blackbirds would appear to be gently falling. They would almost look cute.

But they're not cute. Each "blackbird" has rotating "skeets," like a set of mini hoverboards, that spin to slow down the submunition and allow both laser and infrared sensors to identify the best target, such as a missile turret, tank, fighter jet, just-run car engine, or armored personnel carrier. When the skeet passes an identified target, the software can fire a two-pound penetrator explosive that can rip through heavy-plated armor, ignite fires, and provide a fragmentation ring that throws shrapnel into nearby soldiers. Cluster bombs are released by a person, but after that launch, the sequence of events is fully automated.

Although death and dismemberment from multipronged, multitargeting cluster bombs are extensive, these bombs are more precise than carpet bombing, which is indiscriminate. In theory, CBUs target only sites that match the software. With thermal imaging, a military programmer can even ensure that CBUs avoid the precise temperature range of humans and attack only unoccupied vehicles or equipment. Bombs such as the CBU-97 are far more "future-friendly" in terms of civilians because they are programmed to explode at 50 feet in the air if they cannot find a target. If the sensor fails, a timer activates the bomb, reducing the long-term threat of personnel mines that can kill innocent civilians decades after a war.

However, CBUs are not perfect, and they can kill the innocent. For example, thermal imaging reveals humans in general, not the specific features of soldiers or members of a terrorist group. Also, a bomb may drift and suddenly target a crowded city street or market or a village full of children. A convoy or car containing a target may also carry noncombatants. CBUs may leave unexploded bombs behind—not nearly as often as carpet bombing or sowing mines does, but they do have a failure rate. Children, farmers, anyone taking a casual walk can stumble on an undetonated bomb.

CBUs remain controversial, especially because early versions of the CBU-97 had high rates of safety failure, including estimates of 17 to 20 percent of the bombs remaining live after deployment. In 2008, US defense secretary Robert Gates signed a memo declaring that after 2018 American forces would "only employ cluster munitions containing submunitions that, after arming, do not result in more than 1 percent unexploded ordnance." The requirement, Gates said, could not be waived. He was reacting to the United Nations Convention on Cluster Munitions of 2010, which prohibits the production, use, transfer, and stockpiling of these weapons. Even though 111 nations signed the ban, the United States, Russia, China, Iran, and North Korea did not (a revealing group listing); Russia has since been accused of using cluster munitions in Ukraine against civilian targets. Even with such permissive markets and thus high potential for sales, though, Textron announced in 2016 that it would stop making CBU-97s. In statements in Securities and Exchange Commission filings, the company noted that it had discontinued production because orders had declined and "the current political environment has made it difficult" to obtain sales approvals. However, in November 2017 the Trump administration reversed Gates's 1 percent policy, calling cluster munitions "legitimate weapons with clear military utility . . . providing distinct advantages against a range of threats in the operating environment."

This fluctuating appetite for precision bombing hints at its ongoing challenges, with the most notable being the chance of error. Between 2010 and 2020, the Bureau of Investigative Journalism tracked US drone strikes and other military actions in Pakistan, Afghanistan, Yemen, and Somalia and created a comprehensive list and profile of civilian deaths. In 2020, the group reported its findings to the White House: it counted at least 14,040 confirmed drone strikes, with at least 8,858 people killed (the upper estimate was 16,901). The analysis of this first set of strikes was taken from military records, and thus it was hoped that 99 percent or more of kills in such strikes would be military personnel or targets, not civilians. However, the analysis concluded that at least 910 (10.3 percent) of the drone kills were likely civilians, with an upper estimate as high as 13 percent. The bureau's report found that at least 283, or 3 percent, of the mistaken deaths were of children.

Clearly, these estimated "false positives" exceed Gates's 1 percent limit for error on unexploded munitions and smart weapons. Some have noted, correctly, that the larger fault lies more with the final hard choices made by operators using imperfect information and limited camera footage than with the drones' software and ordinances, but the errors still exist and are still stark.

Almost certainly, the predictive aspects of these tools and intelligence will improve, and it is likely that relevant metadata and information on military versus civilian targets will get better as well. However, for reasons ranging from saving money to the more compelling competitive need to match rivals who can use autonomous decision-making, pressure will increase to replace the humans making decisions with AI, which will open the door to advances that raise a host of issues, including turning war over to so-called killer robots.

Swarming Drones

Today, semiautomated weapons such as cluster bombs seem crude compared with the weapons currently on the drawing boards of militaries around the world.

In his book *Army of None: Autonomous Weapons and the Future of War* (2018), Paul Scharre is blunt. "Artificial intelligence will transform warfare," he writes. "In the early twentieth century, militaries harnessed the industrial revolution to bring tanks, aircraft and machine guns to war, unleashing destruction on an unprecedented scale. Mechanization enabled the creation of machines that were physically stronger and faster than humans, at least for certain tasks. Similarly, the AI revolution is enabling the *cognitization* [Scharre's italics] of machines, creating machines that are smarter and faster than humans for narrow tasks."

More specifically, Scharre discusses how autonomous machines represent the culmination of an arms race in speed. The speed of computer processing has grown as Moore's law has unfolded. Human operators may have a greater sense of context and greater flexibility, but intelligent machines are exponentially quicker to make predictions and act on them. But, as Scharre points out, such abilities immediately raise legal and ethical questions that the US Department of Defense has struggled over. Scharre also notes that *autonomy* is not an absolute term. Smart weapons such as missiles and ICBMs have long had some degree of autonomy. The target and the flight are predetermined, and the missiles can't be recalled—as he writes, it's a matter of "fire and forget"—and they carry advanced guidance systems that can use a variety of technologies to stay on course and seek their predetermined target. But there are missile systems that once launched can use radar to search and select targets; a human operator then approves the attack. One of the problems with smart-weapon systems is that they can easily home in on friendly as well as enemy aircraft or ships. (Early World War II German precision-guided torpedoes had an unsettling tendency to circle back and attack the U-boat that launched them.)

Full autonomy is something entirely different. In the essay "The Ethics & Morality of Robotic Warfare" (2016), the University of Pennsylvania political science professor Michael Horowitz succinctly defines an autonomous weapon as "a weapon system, not a person, that selects and engages targets on its own." A fully autonomous weapon would not only search for and identify targets but navigate what could be a complex, treacherous battleground and—this is where the ethical questions emerge—make the decision to attack and kill or not. Human operators might play little or no role, even in determining targets. Autonomous drones, for instance, could hover, waiting to detect a target.

Thus, there is a self-fulfilling prophecy of lost control with respect to these systems. In an attack, for example, only autonomous drones might be able to react quickly enough to be effective, and they could accelerate the action on the battlefield to the speed with which a computer could calculate all the probabilities. They have the potential to drive the action on a coordinated, multidimensional battlefield network of significant size to levels of complexity and speed that are simply beyond the full comprehension of human operators. Like all weapons systems, they drive developments with a dynamic that is nakedly bottom line: who wins and who loses, who lives and who dies.

That may be why there has been such resistance in the Department of Defense to full autonomy. As Max Tegmark, a physics professor and AI researcher at MIT and the president of the Future of Life Institute, wrote in an open letter coauthored with University of California computer science professor Stuart Russell in 2016, "If any major military power pushes ahead with AI weapons development, a global arms race is virtually inevitable, and the endpoint of this technological trajectory is obvious: autonomous weaponry will become the Kalashnikovs of tomorrow," referring to the global ubiquity of the Russian assault rifle. The letter noted how relatively inexpensive such weapons are—certainly compared with nuclear arms— and how easily they could spread. "It will be only a matter of time before they appear on the black market and in the hands of terrorists, dictators wishing to better control their population, warlords wishing to perpetrate ethnic cleansing."

Scharre focuses on development work undertaken on another aspect of drone warfare: swarming drones. Right now, one of the cutting edges of military autonomy is the ability to control not just one small drone but a swarm of them. But *control* may be the wrong word. In fact, the underlying process here may more resemble the kind of interplay between a bird and its flock. Birds in a flock, like ants in a colony, appear to operate seamlessly and with purpose because they are wired to quickly respond to a handful of relatively simple stimuli. For birds, that may involve flying at a certain distance from the other members of the flock or following the bird directly in front of them. The flock operates autonomously, with no central controller. Now take this to another level: airborne combat. Scharre describes a relatively simple algorithm, called the "greedy shooter," that shapes the behavior of two drone swarms engaged in aerial combat. The swarms are autonomous: after a human operator hits a button to start, the attack develops without human intervention. In fact, given enough data, the attacks could be simulated ahead of time, which is a powerful tool of prediction.

Superficially, there is an element of play to all this. The drones themselves are unarmed—they "win" with accurate simulated shots—but, even so, the aerial combat creates a fast, swirling, intricate battle. The big question is not just how effective these algorithms are as weapons but, rather, what will happen when powerful AI is programmed into more powerful computing devices on the drones. And more than just drones can be programmed in this way. Why can't robotic ships or tanks swarm? Recall the problems of unexploded bombs scattered across a war-torn countryside. In Iraq and Afghanistan, the United States found that the best way to clean them up was with simple ground-based robots known as PackBots—more than 6,000 of them by 2015. But what if these robots carried their own bombs and were drawn to human targets?

AI would allow drones not only to engage in complex tasks but also possibly to *learn*. In terms of prediction, AI systems such as neural nets could turn drones or missiles or entire networked battlefields into chesslike games that could quickly evolve in ways incomprehensible to their human observers and ostensible masters. Success in warfare has long been a matter of better weaponry and intelligent tactics and strategy: technology and prediction. AI warfare will become ever more dependent on blindingly fast, accurate prediction—a military version of high-frequency trading. And that opens the door to a variety of new risks.

AI and Ethics

Let's begin with the ethical questions that arise from the development of autonomous weapons systems—what some call "killer robots." There are a number of considerations, some secondary, some fundamental, some ethical, some legal. Machines—at least, any we can imagine today—may be fast and farseeing, but they lack humans' ability to discriminate and to understand, even if fallibly, and to examine complex context. For example, a human can readily discern that a stick is not a gun, that a soldier is wounded, that a soldier is surrendering. Machines so far are not so discriminating.

Legally, autonomous weaponry may violate notions of a just war, notably *jus in bello*—that is, the conduct of war—or transgress what's known as the Martens Clause, which was formulated in 1899 at the Hague Convention. Friedrich Martens, a Russian diplomat, set out a rule that applies to individuals caught up in a war where no international law applies: "Human persons are protected by the principles of humanity and the dictates of human conscience." The more fundamental rule is that humans are accountable because they have consciences (some perhaps more than others) and can be punished for violating the maddeningly vague "principles of humanity."

The Martens Clause has long been subject to argument and reinterpretation, particularly over the use of nuclear weapons. In the case of autonomous weaponry, it's true that a machine, no matter how intelligent, cannot reflect on goals and purpose, right and wrong. Punishment is meaningless. A machine, in short, lacks human agency and long-term responsibility. An autonomous killing machine is thus, almost by definition, amoral.

What's clear are two realities: Militaries around the world possess and use semiautonomous weapons such as human-piloted drones, but even countries such as the United States and China, which are fully capable of developing autonomous weaponry and have engaged in research on it, have held back on actively deploying it. In 2020, after 15 months of deliberation, the US Department of Defense officially adopted a series of five ethical principles for the use of AI. The agency concluded that "AI technology will change much about the battlefield of the future," noting that "the adoption of AI ethical principles will enhance the department's commitment to upholding the highest ethical standards . . . [and] embrac[e] the U.S. military's strong history of applying rigorous testing and fielding standards for technology innovations."

The five principles are that the development, deployment, and use of AI capabilities must be (1) *responsible*, with appropriate levels of judgment and care; (2) *equitable*, taking deliberate steps to minimize unintended bias in AI capabilities; (3) *traceable*, meaning that AI will be developed and deployed so that the personnel managing and using the systems possess an appropriate understanding of the technology, development processes, and operational methods applicable to AI capabilities, including auditable methodologies, data sources, and design procedure and documentation; (4) *reliable*, with explicit, well-defined uses and with the safety, security, and effectiveness of such capabilities subject to testing and assurance within those defined uses across their entire life cycles; and (5) *governable*, with AI capabilities designed not only to fulfill their intended functions but also to possess the ability to detect and avoid unintended consequences and the ability to disengage or deactivate deployed systems that demonstrate unintended behavior.

Though such principles sound reasonable, we are in uncharted territory. The laudable goal of having AI be "traceable" so that personnel will "possess an appropriate understanding of the technology" may be harder than it looks. A 2017 study from Pegasystems titled "What Consumers Really Think

about AI" showed that while 34 percent of people thought they had inter-
acted with AI, 32 percent were not sure. The truth was that 84 percent had
done so. This implies that half of the people using AI don't even know it.

Beyond this traceability fog, it's not clear how well these guidelines will
translate into real-world combat or how to avoid the biases of a program
that is deciding who is a target and who is not (the reliability problem). The
bias of the machines will *always* reflect the bias of the programmers who
make them.

Five Levels of AI

Autonomous weaponry taps deeper concerns about artificial intelligence.
On August 2, 2014, Tesla CEO Elon Musk famously tweeted: "We need to
be super careful with AI. Potentially more dangerous than nukes." Musk's
tweet was based on the presumption that eventually, as he said on CNBC in
July 2017, "robots will be able to do everything better than us" and that a
self-aware intelligence may someday emerge that is a "fundamental risk to
the existence of human civilization." Because Musk is actively developing
AI systems for self-driving cars at Tesla and is a cofounder of Neuralink, a
company working to create a way to connect the brain with machine intelligence, his comments stirred anxieties. Some immediately criticized him,
noting that fears of machines turning on humans like in a *Terminator* movie
are unfounded, given the AI technologies that exist today or will exist in
the near future. Some, such as Facebook cofounder Mark Zuckerberg, even
called such statements irresponsible.

But Musk, in an April 6, 2018, tweet, clarified that his fear was not about
a narrow-task AI, like the kind that enables self-driving cars, but about a
more complex and self-aware "digital super-intelligence" that could at first
take jobs and then, eventually, threaten humanity itself—a sentiment sup-
ported by the former Google CEO Eric Schmidt and Toby Ord in his book
The Precipice (2020), among others. Musk went on to participate in and
fund efforts by the AI community, including Max Tegmark's Future of Life
Institute, to build safety into AI processes. Still, Musk's questions remain.
Could AI really attain a state that could threaten humanity? Could autonomous weaponry, powered by AI, be the cutting edge of that eventuality?

Perhaps, but first we need to break down AI into four kinds of machines
based not on structure or function, such as deep learning or NLP, but

more on abstract attributes with increasing complexity: (1) ability to react, (2) limited memory, (3) theory of mind, and (4) self-awareness. Reactive machines lack any concept of the past, so they have no working memory of what happened, say, in previous games and have no way to construct the future. Deep Blue and AlphaGo, the AI that respectively mastered *Jeopardy!* and Go, were reactive; they could not adapt easily or at all to simple rule changes, such as adding a new piece.

One of the most powerful examples of limited-memory AI that can look into the past is the self-driving car. This kind of AI can observe other vehicles' speed, movement, and direction. Such computation requires a working memory and the ability to track specific objects, monitor them over time, and predict trajectories. These observations and predictions merge with knowledge of traffic laws, signals, data from the car, global-positioning data, and other user inputs. However, the self-driving car's knowledge of the past is limited. The software cannot accumulate its own experience and learn like a human driver, read the face of another driver to derive intention, or recall complex situations. But eventually it will.

The third type of AI would be able not only to see, quantify, and make predictions about the world from its own data but also to make inferences and predictions. In psychiatry, this is called having a "theory of mind," where you become aware that other conscious entities have thoughts, emotions, volition, and direction. Most important, AI would project this awareness and respond to it as most humans do without any effort in conversation, where we read body language, tone, eye direction, and unspoken communication to infer others' responses.

Of note, not every human even has this ability. Autism spectrum disorder (ASD) is a complex psychiatric disorder in which individuals may lack such a theory of mind and thus cannot understand why some of their actions might upset others; this inability is why people with ASD can seem distant and unengaged to people without ASD. It is possible and even likely that future machines might also have a range of cognitive abilities; indeed, Deep Blue became Deeper Blue, more advanced versions of AlphaGo have been developed, and the AI of the future will almost certainly feature a variety of attributes. Even human individuals evolve versions of themselves as they age and change.

The fourth state of AI is a machine that can form a representation of itself and view it from the outside—a form of self-consciousness. This would become the final metamorphosis of Descartes's aphorism "I think, therefore I am" but taken to the self-reflective step where a machine could state, "I think about me thinking that I am." At this point, the machine might come to realize its own mortality, become self-defensive, and seek self-preservation.

This is a common theme in popular science-fiction movies such as *Terminator* (1984), *The Matrix* (1999), *I, Robot* (2004), and others in which an AI system decides that humans pose too large of a risk to its own survival and that it must exterminate us all.

Even though humanity has a spotty track record of preserving species, we do have a key, unique trait that represents hope for a new AI. Humans have evolved in such a way that our self-consciousness places us at what can be described as an even higher, fifth level of AI: extinction awareness. We are aware of the actions, trajectories, and changes that affect our species as well as all other species on the planet. We found and fixed a hole in the ozone layer, and we are tracking changes to Earth's climate. Interspecies and planetary-scale engineering transcends the life span of a person or even a family; it is the stewardship of life itself, including AI, which could become a form of life.

Given that humans eventually acquired and have acted on extinction awareness to preserve life, there is a possibility that a self-aware AI could attain this fifth stage of intelligence and decide that humans present not only a partial risk, but also a possible benefit, to its long-term survival. This future AI may appreciate that the viability of humans, other creatures, and varied types of AI are important for its own self-preservation. This, however, depends not only on our own programming of nascent AI but also at some future time on AI's ability to program itself. But how well can AI be programmed for morals, and is there a set of universal morals that we can teach machines? It turns out, we know some of the answer already.

The Moral Machine

In 2018, the researchers Azim Shariff, Jean-François Bonnefon, Iyad Rahwan, and their colleagues published "The Moral Machine Experiment," a paper that stemmed from an online experimental platform exploring moral dilemmas faced by autonomous vehicles and that used data on 39.61 million decisions, spanning ten languages, by millions of people in 233

countries. Participants had to decide what to do if a self-driving car suffered brake failure: continue straight, which would kill pedestrians but save
the passengers, or swerve, saving pedestrians but killing passengers. There
were only two possible outcomes, and each would result in death. The
study then presented variables of this scenario across nine factors: sparing
humans (versus pets), staying on course (versus swerving), sparing passengers (versus pedestrians), sparing more lives (versus fewer lives), sparing
men (versus women), sparing the young (versus the elderly), sparing pedestrians who were crossing legally (versus jaywalkers), sparing the fit (versus
the less fit), and sparing those with higher social status (versus lower social
status). Additional characters appeared in some scenarios, such as criminals, pregnant women, and doctors, who were not linked to any of the nine
factors but existed merely to gauge their impact.

The results were a fascinating exploration of moral choice under pressure. First, the researchers found three distinct "moral clusters" that were
broadly consistent in both geographical and cultural proximity: Western,
including North America and European countries that belong to Protestant,
Catholic, and Orthodox Christian cultural groups; Eastern and Far Eastern,
including countries such as Japan and Taiwan of the Confucianist cultural
group and countries such as Indonesia, Pakistan, and Saudi Arabia of the
Islamic cultural group; and Southern, including Latin American countries
as well as countries with some French influence, such as French overseas
territories.

The differences among the three were striking. The authors observed
"systematic differences between individualistic cultures and collectivist
cultures." The preference for sparing the young (versus older people) and
those of higher economic status was much less pronounced in the Eastern
cluster and much higher in the Southern cluster. Countries in the Southern cluster exhibited a much weaker preference for sparing humans over
pets, compared with the Eastern and Western clusters. Individualistic cultures, defined as those that emphasize the distinctive value of individuals, showed a stronger preference for sparing the greater number of people.
Participants from collectivistic cultures, defined as those that emphasize
respect for older members of the community, showed a weaker preference
for sparing younger people, as expected.

Some intercluster differences revealed that individuals were preferred for
demographic and country-specific reasons. For example, the authors found

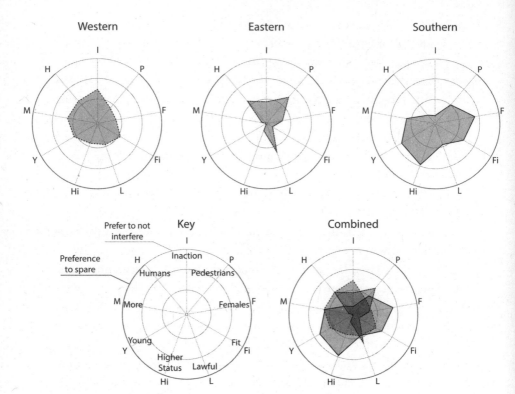

Figure 8.1
Clusters of moral codes between countries. One hundred and thirty countries with
at least 100 respondents were selected (range, 101–448,125). Distributions across
the three clusters (*top row*) reveal stark differences between parts of the world. For
instance, the Eastern cluster (Eastern hemisphere countries) consists mostly of countries of Islamic and Confucian cultures. By contrast, the Western cluster (Western
hemisphere countries) has large percentages of Protestant, Catholic, and Orthodox
countries in Europe. Countries belonging to the Southern cluster (Southern hemisphere countries) show a strong preference for sparing females compared to countries in other clusters. Combined clusters (lower right) and moral decision points
(lower left) are also shown (*bottom row*). *Source*: Adapted from Edmond Awad, Sohan
Dsouza, Richard Kim, Jonathan Schulz, Joseph Henrich, Azim Shariff, Jean-François
Bonnefon, and Iyad Rahwan, "The Moral Machine Experiment," *Nature* 563, no.
7729 (November 2018): 59–64.

that higher country-level economic inequality, as calculated by the country's Gini coefficient, corresponded to how unequally individuals of different social status were treated. Countries with more equality between rich and poor treated the two groups more equally in a life-or-death decision.

Sex was also a dividing line. The study revealed a large country-level gender gap in the health and survival of women by calculating the ratios of female-to-male life-expectancy and sex ratios at birth (a marker of female infanticide and antifemale sex-selective abortion). In nearly all countries across all clusters, participants showed a preference for females, yet this preference was stronger in nations with better health and survival prospects for women. The authors wrote that in countries where there is "less devaluation of women's lives in health and at birth, males are seen as more expendable."

Yet some trends were universal. A slight preference for sparing pedestrians over passengers, and for sparing the lawful over the unlawful, appeared to be shared by all cultures in all data clusters. The overall preference for saving women versus men mostly held. But in general, an ideal set of universal ethical laws for machines may not yet exist across the world's ethics.

Nevertheless—though this complex task is well beyond our current capabilities—AI could *possibly* be programmed to match the needs of the society that built it and to create predictions based on nuances that change when a machine crosses a border into a new "moral" cluster. Indeed, this means that country-specific normative ethics could at least be considered for, if not implemented into, frameworks that would be making life-and-death decisions. This implementation could eventually lead to completely autonomous machines that carry an "ethical chip," making them similar to the continuously adapting, imperfect humans who programmed them.

Military Risks

There are darker possibilities, of course.

As a matter of prediction, we can barely see past the next generation of AI-enabled weaponry; the fifth level, whether it's a Musk-forecasted extinction of humans by machine or a flowering of moral machines that defend life, is not yet clear. What exists directly before us is unsettling enough. The dynamics of war hinge on shifting advantages, often seemingly minor, of offensive tactics versus defensive ones, which are increasingly fluid, as we see in the global, highly technical, and shadowy jousting of cyberwarfare. AI accelerates both the predictive quality of autonomous weaponry and the rapidity of adjustment and modification, which resembles aspects of the nuclear arms race (and follows earlier destabilizing arms races, such as battleships), but at a potentially much greater rate of change. And, of course, the new technologies do not exist in isolation. They join bulging weapons arsenals that have already been amassed all around the world. Imagine strapping a nuclear weapon onto an autonomous drone or chemical weapons onto swarms of miniaturized drones.

Warfare is one area where enhanced prediction by autonomous weaponry *increases* risk rather than reduces it. The shadow of military risk is a shroud as dark as the night—and it's been lengthening with the sequential development of automatic weapons, air power, tanks and missiles, chemical warfare and nuclear bombs, AI, and cyberwarfare. Weapons technology has grown exponentially more powerful, more rapid, more surgically delivered, and, of course, increasingly intelligent and predictive. This mix is lethal. The ability of weapons to calculate, choose, and learn with blinding speed raises the potential for disaster to much higher levels. Weapons are increasingly precise and more destructive; they can be employed by nations with enormous resources or by insurgents and terrorists pursuing asymmetric warfare. Long before the technology exists to attain the fifth stage of AI or for machines to develop a sense of self-consciousness, we could easily destroy ourselves with weapons that we dispatch, if not control—weapons that are both unthinking and intelligent.

We, in short, are the problem.

This enormous risk of AI-enabled warfare arguably plays a larger role in the human reaction to autonomous, predictive-based weaponry than do moral or ethical considerations, which, as history shows, can be easily shredded, twisted, and tossed aside. That fear of intelligent, predictive weaponry is responsible for nations keeping true autonomous weapons on the shelf so far (or, for that matter, not using nuclear weapons, though it hasn't stopped proliferation); for statements such as the US Defense Department's AI principles; and for attempts by groups such as Tegmark's Future of Life Institute—the name says it all—to get responsible control over AI development. Is that possible?

Maybe. Such control would involve a long-range prediction planted deeply in the realm of global and species-wide uncertainty. If history is any guide, humanity has a strong penchant for war and for using the brightest new toys of destruction available. The odds of predicting war, sadly, have always been greater than those of predicting peace. Despite such a gruesome history, in the past few decades humanity has experienced a downward trend in the number of deaths from war, among the lowest shares of battle-related deaths in history, according to Harvard University cognitive psychology professor Steven Pinker. Against these trends, the scale of destruction in recent wars (Syria, Ukraine) is a reminder of a sobering reality.

9 Predicting Performance

Every year millions of people apply for jobs: big jobs, little jobs; full-time or part-time jobs; blue-collar or white-collar jobs; internships. This process has been going on for decades, but the pandemic (with its flood of workers quitting current jobs to find better ones) has helped move job hunting onto computers, and new algorithms have changed the game entirely. Job hunting has become a game much like social media, with job seekers desperately shaping themselves in ways they think will be attractive, while companies and vendors press ever harder to automate and seek ever more predictive tools that can anticipate how candidates will function when they get the job.

As a result, job hunting has become more and more a mystery for both ends of the process. It involves the writing of a résumé, the composing of a letter hardly anyone reads (except an algorithm hunting for keywords); the filling out of applications; the hunt for references; the anxiety of the interview, often an asynchronous video interview without a human interviewer; and a battery of tests. A large corporate bureaucracy has grown up around hiring, consisting of consultants, experts, recruiters, human-resource staff, test developers, and, increasingly, model builders. For job seekers, the process is an anxiety-provoking puzzle that has spilled over well beyond the job hunt. Admission to colleges and universities now features its own armies of consultants and admissions experts, tests, interviews, essays, and an ever-expanding demand for credentials—as well as, famously, charges of bias and cases of corruption, both civil and criminal. In both job hunts and college admissions, the challenges are wildly inflationary: job seekers send out reams of résumés; college seekers dispatch dozens of applications. That process has also now seeped down to elite and expensive preschools, which even test, and deeply interview, three-, four-, and five-year-olds.

This is all a mass exercise in trying to define and capture *talent*—in some cases even before it has a chance to blossom. Thus, the hunt is also about a person's potential. Yet when the candidates get older, the view is predominantly backward looking and static. Companies claim to want the best talent they can find; schools, from high to low, say they want students who will not only perform at an elite level but go on to elite jobs. It's no shock that the accumulation of credentials—degrees, awards, quantifiable assets—tends to be self-fulfilling and endlessly self-assessing.

The Challenges of Predicting Success

There is, however, a problem with this process: we're not very good at assessing future performance. We have never been all that good at it, and some claim that we are getting worse at it. As Peter Cappelli, a management professor at the University of Pennsylvania's Wharton School, notes in a *Harvard Business Review* survey of the hiring scene: "Businesses have never done as much hiring as they do today. They've never spent as much money doing it. And they've never done a worse job of it."

Cappelli's critique is sweeping. A great deal changed in the way hiring took place from the post–World War II period to the 1970s, when 90 percent of hires came from within the same company. The majority of hires today are not people searching for jobs but so-called passive candidates, most of whom don't realize they've become active candidates. Companies, Cappelli writes, try to fill a "recruiting funnel" with as many names as possible, so they're constantly searching for new recruits, even if they don't have jobs to fill. (In some cases, actively looking for a job may be a disadvantage.) According to the consulting firm Korn Ferry, about 40 percent of companies have outsourced hiring; they use subcontractors to search the internet for candidates and track key words. Other vendors specialize in, as Cappelli says, "an astonishing array of smart-sounding tools that claim to predict who will be a good hire."

We often say that finding and hiring the best possible talent is key to driving our labs and businesses. Many CEOs say hiring is their biggest challenge. Indeed, many of the arguments for AI-enabled recruiting begin with the notion that human capital is a much more important asset than ever before. Yet within companies hiring is a cost center that's not afforded prestige or resources. Rising turnover is a problem that has transformed hiring from a custom task into an industrial process driven by automation. Hiring takes a relatively long time and is expensive; it costs a company an average of $4,129 to hire someone, notes Cappelli. And then, who has the company got? How long will they stay, and what will they do?

Hiring is always an attempt to *predict* human behavior and future performance. This task is far more complex than key words on a résumé written by a consultant because measuring loyalty, hard work, and competence can be difficult. Subject-matter expertise is somewhat easy to gauge because it is based on education and work experience, but it is not as easy to measure a willingness to learn new things or innate ambition. Also, it is hard to predict how new hires will fit into a company's culture, or if they will help lead the company to new heights versus create disruptions. These challenges happen with every new hire.

The paradox here is that, although CEOs say hiring is the most important thing they do, many companies are disengaged from the process. Only a third of US companies say they monitor whether their recruiting process system is working or is cost effective. One unintended consequence of hiring from the outside is that it allows companies to reduce training, development, or assessment of employees, which means constant hiring beyond the traditional entry level. Retention has all but collapsed. Census data suggest that 95 percent of hiring before the COVID-19 pandemic was to fill existing positions—Cappelli was writing in 2019—at a time when millions were not suddenly quitting or retiring. In fact, as Cappelli notes, hiring is increasingly not about the ability to master new areas and skills but about doing a specific job. For that, knowing someone's recent job experience is all you need.

Companies aren't alone in struggling with performance assessments. A pair of studies by a research team at the University of North Carolina at Chapel Hill and by another at Vanderbilt in 2017 showed that supposedly "objective metrics" for accepting graduate students to competitive PhD programs in the United States did a poor job of predicting who would be successful. The North Carolina study noted that "standardized test scores, undergraduate grade point average, letters of recommendation, a resume and/or personal statement highlighting relevant research or professional experience, and feedback from interviews with training faculty" should, in theory, correlate with research success in graduate school. Yet none of these metrics appeared to matter. Indeed, not one of the conventional "objective credentials" could predict the normal metrics of early-stage scientific success, such as the number of first-author publications or conference presentations, the ability to acquire fellowships or grants, graduation success, qualifying-exam success, or time to degree. Moreover, the Vanderbilt data set showed that Graduate Record Examinations (GRE) scores were only "moderate predictors of first semester grades" and "weak to moderate predictors of graduate GPA." The authors concluded even more strongly that there is "no relationship between general GRE scores and graduate student success in biomedical research."

Although standard metrics for predicting success struggled, according to these studies, that does not mean they didn't provide clues for more fertile areas of inquiry. The feature with the greatest capacity to predict future success turned out to be letters of recommendation from undergraduate teachers and scientific mentors. Unexpectedly, traditional *subjective* assessments from people who knew the students for the longest period of time in the context of their chosen field provided the most relevant metrics.

Using predictive algorithms to mine letters of recommendation could thus become a new metric for admissions. Such free-text and predictive analyses are an important part of NLP algorithms, which scan and organize large amounts of text to look for patterns. By converting the text into vectors or abstract data structures, new patterns may be found, quantified, and analyzed relative to other metadata in order to create new predictions.

Prediction Games

In fact, predictive algorithms of many sorts have swarmed into the hiring game, though so far there's little evidence that they have transformed outcomes. The claims for AI-enabled recruiting are large. In one survey of the field in 2019, the authors J. Stewart Black and Patrick van Esch argued that we are at the start of what they called "Digital Recruiting 3.0," driven by the shift to human capital and the increased use of machine learning. (Recruiting 1.0 and 2.0 involved advances in search and data.) "Computers can now perform tasks and make decisions that normally require human intelligence," they wrote. "Some key potential advantages include the ability to more effectively identify, attract, screen, assess, interview, and coordinate with job candidates." AI can process information and make decisions at volumes and speeds far beyond what people can do, the authors noted.

AI-enabled recruiting tools and systems can "overcome common cognitive biases that hurt the reliability and validity of human judgment in recruiting activities."

Many of the new algorithmic tools are, by any measure, straightforward and effective. They gather, screen, and coordinate many candidates, usually on the internet. Although data gathering itself is not a predictive activity, the data assessment and modeling are very much a prediction game. Machine learning is often used to gather a huge pile of names, eliminate 50 to 80 percent of the candidates, and then cull the remaining pool. Unilever, for instance, used a series of 12 neuroscience-based test games from a company called Pymetrics on internship candidates who had survived a first screening. One three-minute game attempted to assess "risk-taking" by inflating a balloon with money and balancing the opportunity to score high on the money front with the risk that the balloon would burst. As the authors of the survey in 2019 admitted, before the game could be designed, Unilever needed "to understand the relationship between risk propensity and job success for certain positions." Company research showed, not surprisingly, that moderately high levels of propensity for risk were positively related to job performance as opposed to very low or very high propensities. In theory, the company could supplement this result by doing a genome test and checking the applicant's dopamine genes (though, as we have and will see, that step could raise privacy and legal issues).

After applicants played the games, a third of them moved on to the asynchronous video interviews. The questions were based on the capabilities and characteristics of "successful and average" employees, in this case interns. The AI system analyzed not just the answers but also word choice, tone of voice, and microfacial movements. Based on this research, Unilever determined what capabilities and characteristics were likely to lead to success. The vendor, HireVue, later removed the microfacial part of the assessment due to privacy concerns.

Experts on hiring admit that, for many decades, the hiring process has been biased in favor of men, whites, graduates of a small number of schools, and employees who resemble corporate decision makers—who "fit in" with what society already looked like. Machine learning, which is often trained on data from a company's hiring and reviewing history, has struggled to escape those tendencies, although its proponents often insist that it is free of cognitive biases. Such traditional practices sparked legislation such as Title VII of the Civil Rights Act of 1964, which bans discrimination by race, color, national origin, sex, and religion in hiring but is difficult to enforce, and using proxy data such as elite schools, family backgrounds, or even proficiency in certain sports (golf, lacrosse) only replicates deep systemic biases.

Unilever did conduct some research to try to define "successful" employees, though whether that research could predict a reasonable percentage of candidates who would thrive remains a more difficult question. Even if it could, why would the same system of prediction necessarily work for other companies or other jobs? The use of test games also creates a situation in which candidates learn over time how to game the games. Hiring has always had this aspect: candidates have long been urged to try to understand what a prospective employer wants and to provide it—by, for instance, mastering interview questions beforehand. Various so-called psychometric tests have been used for nearly a century, though they have been limited since the passage of the Americans with Disabilities Act in 1990, which forbids employers from inquiring about physical or mental disabilities. And, of course, an intense debate over assessment testing in general—SATs, GREs, IQ tests, and a variety of personality tests—has simmered for decades.

In a wide-ranging 2019 *Harvard Business Review* article looking at the legal and ethical issues around using AI in hiring, Ben Dattner and coauthors note that there is little information available about many of the new AI assessment tools, which are "technological innovations, rather than scientifically derived methods or research programs." As a result, they conclude, "it is not always clear what [these tools] assess, whether their underlying hypotheses are valid, or why they may be expected to predict job candidates' performance."

Predicting Athletes and Twin Astronauts

There is one business that's often considered a paragon of talent scouting, recruitment, and hiring: professional sports. Few professionals receive as much scrutiny over a long period of time as athletes. They are scouted, filmed, tested, poked, and prodded from childhood on. How fast do they run? How high do they jump? How much can they lift? Sports, particularly at the highest levels, claim to be among the most meritocratic of activities: it comes down to performing in visible ways—winning and losing.

The recent trend toward applying massive amounts of data and analytics to athletics is an effort to predict performance and outcome that's not all that different from forecasting securities prices or genomic tendencies; the surge in legal sports betting adds another layer of analysis and prediction. But for all the time and effort put into attempting to select the very best performers and maximize their performances through increasing amounts of data, even casual sports fans know of highly touted young athletes who flame out and relative unknowns who end up giving hall-of-fame speeches.

Much of such outcomes, of course, turns on aspects of high performance that resist quantification. There is much we still don't know about biology or about the relationship between the body and the mind, as well as the relationship between nature and nurture. The application of tools such as machine learning to performance and proxy data may improve the ability to predict, say, successful draft picks, discounting the effect of uncertainties such as bad luck, bad coaching, personal difficulties, and injuries. But the kind of data that may take some uncertainty out of scouting and assessment may involve what's normally viewed as invasions of privacy, which may create a variety of data limitations and may run the risk of overfitting—that is, finding a false pattern in inadequate data.

Let's start with the gene and with the conventional wisdom that genetics is part of your destiny.

If you are an astronaut on the International Space Station, your genome is an open book: astronaut DNA gets tested, sequenced, matched, and monitored. Indeed, all the cells that astronauts shed on the walls and in the air of the space station mix with the DNA remnants of previous inhabitants. In 2015, Chris became a principal investigator of the NASA Twins Study, analyzing the human and microbial DNA of astronauts and identical twins Scott and Mark Kelly. During the study, the team compared the Kellys' DNA to data collected in previous years. Both astronauts showed some evidence of the "ingestion" of microbes found on the space station walls. Scott, who has spent nearly ten times the number of days in space as Mark, showed that 56 percent of his gut microbes matched the gut microbes of others on the space station, whereas Mark showed 5 percent less. To understand any change during spaceflight, we use all the data from the environment, the astronaut, ground simulations, and even other organisms such as flies, worms, rodents, and bacteria to build a better model (figure 9.1).

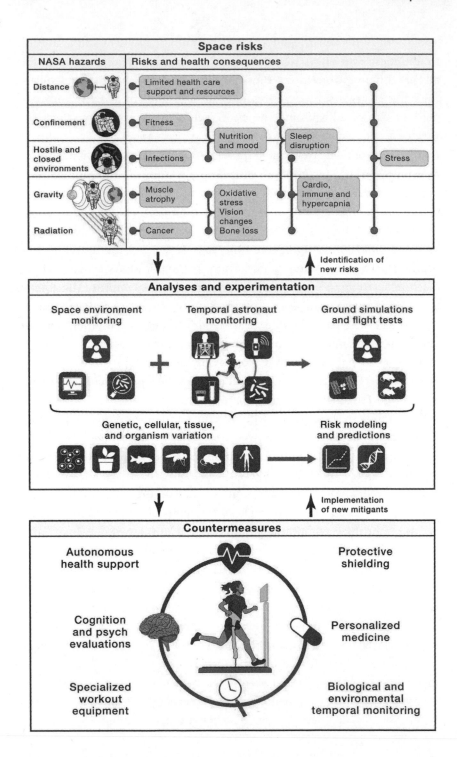

Beyond the microbial species, Chris and his fellow investigators detected intriguing changes in human genes from time spent in outer space. During the analysis of gene-expression data, it became clear that some of the genes that had turned on while Scott was in space for a year did not completely turn off when he returned to Earth. They included genes for DNA repair and immune-system maintenance as well as for bone and muscle formation. It was obvious from looking at the genetic data that Scott's body was still adapting to Earth's gravity after he returned home. The same changes did not occur with Mark, who had remained on Earth, so it seemed as if these changes were unique adaptations of activated genes that were a consequence of spaceflight.

Gene expression is not the same as the catalog of genes in a person; gene expression creates a level of greater complexity, flexibility, and potential. Because the Kelly brothers are slightly different due to varying exposures to space and, with the passing of time, greater genetic divergences, would it be possible to pick one as a better choice to send into space? Are there genes that we could add, tweak, or modify that would make humans better suited for spaceflight—or for, say, accounting? Could we predict, based on someone's genetic code, who would be best for a given job? Could we use knowledge of individual genomes to make job-related assessments? *Should we?*

Predicting which genes will create specific traits is an age-old dream, pondered by every parent, that straddles the difference between nurture (the effect of the environment) and nature (your genetic inheritance). To know what genes are important for a specific phenotype, you first need to know if that trait is driven by training and experience or innate ability and genetics. Almost all of our physical and mental traits have some

---

Figure 9.1
Predicting astronaut health. NASA and other space agencies build data models for their astronauts based on medical data, environmental monitoring, model organism testing, previous experiments, and many other factors to create predictive models and develop countermeasures to mitigate the risks. *Source*: Sonia Iosim, Matthew MacKay, Craig Westover, and Christopher E. Mason, "Translating Current Biomedical Therapies for Long Duration, Deep Space Missions," *Precision Clinical Medicine* (December 2019): 259–269.

genetic basis; it's just a question of how much. From previously published studies of identical twins, such as those by the Dutch biological psychologist Dorret Boomsma and colleagues, religion is one of the few traits that have no genetic component (a good negative control). Other traits, such as height, eye color, and susceptibility to a range of diseases, are more obviously related to genes.

Much of the estimated heritability for traits comes from twin studies, in which researchers examine the difference in a given phenotype for identical (monozygotic, or MZ) twins, such as Scott and Mark Kelly, who were born with the same DNA, and dizygotic (DZ) twins, who have 50 percent of the same DNA and shared the same womb. The larger the difference between the twins for a given trait, the more that trait can be attributed to the environment. This difference is called the "heritability estimate," or h2, which is twice the difference between MZ and DZ correlations: h2 = 2(pMZ – pDZ).

For example, typical MZ and DZ correlations for depression are about 0.4 and 0.2, respectively, and heritability is estimated at around 40 percent (h2 = 2(0.4–0.2)). Higher correlations are usually observed for lifestyle factors, indicating the importance of a shared family environment. For the likelihood of smoking during adolescence, typical MZ and DZ correlations are 0.9 and 0.7, respectively, also leading to a heritability estimate of 40 percent.

Other traits with very high heritability include specific molecular phenotypes such as lipoprotein (cholesterol and triglyceride) levels (90 percent) as well as the likelihood of certain behaviors, which may be surprising. For example, boredom susceptibility, although seemingly a function of age or access to distracting technology, also has a genetic component of 52 percent, as indicated by twin studies. Yet this is not as high as the heritability estimates for anxiety (58 percent), disinhibition (61 percent), or even adventure-seeking behavior (65 percent). Such estimates raise the question: How much of our personality is truly our own compared with a subtle influence of traits distilled over the coals of our ancestors' sexual choices?

The Right Genes for the Right Job

Given heritability estimates for so many traits, it might seem that, with enough data, the likelihood for success in a specific job *could* be successfully predicted. If that were the case, we might find ourselves, depending on our point of view, either in an occupational utopia that would place people in the right job with the right coworkers, or in a dark, vocational dystopia in which lives are rigidly determined by genetics.

This scenario raises the always controversial question of genetic engineering for performance.

A wealth of genes have been found by studying outlier and extreme populations, such as mountain climbers and people who should have a disease such as AIDS yet have avoided it. In the large-scale Resilience Project at the Icahn School of Medicine at Mount Sinai in New York City, there is an ongoing effort to find people who should be dead but are not: genetic "superheroes" who have mutations that should lead to a significant disease but for some reason don't.

The project has already found new classes of superheroes for HIV resistance. One group is made up of so-called elite controllers—patients who maintain HIV at very low levels even though they remain infected; another consists of long-term nonprogressors, those who never lose their immune-system function; and then there are the elite neutralizers, who produce highly potent, neutralizing antibodies to HIV that are not normally seen. In 2011, building on such discoveries, Harvard Medical School's George Church proposed a list of protective genes that are known to confer advantages on their lucky hosts and could be used to stratify potentially high-risk and low-risk employees for specific jobs.

Those hiring for high-stress environments should look no further than PDE4B: higher expression of this gene can be associated with lower anxiety and higher problem-solving capacity. Higher expression of the APOE, TERT, and APP genes appears to provide physical and mental longevity, and higher levels of the DEC2 gene allow humans to sleep less but still function at high levels and achieve increased overall vitality. Employees with such genes could be encouraged to apply for jobs that require long hours, such as physicians, drivers, and pilots, and their gene-expression patterns could be monitored to make sure the genes are fully genetically active.

Overexpression in the so-called sirtuin genes—SIRT1, SIRT6, and SIRT7—has been implicated in influencing a wide range of cellular processes, including aging, transcription, apoptosis (a kind of programmed cell death), inflammation, and stress resistance as well as energy efficiency and alertness in low-calorie situations. Sirtuins can also control circadian clocks and mitochondrial biogenesis (in which cells increase mitochondrial mass) and help people get back to work after long shifts.

If you were planning to travel in space with limited oxygen or climb mountains like Sherpas in the Himalayas, a well-developed cardiovascular system would be necessary to increase oxygen uptake, maintain hemoglobin function, and secure normal blood levels. It could also provide protection against increased cardiovascular disease risk. Studies show that Tibetans have variants in the EPAS1 gene that provide larger chest and lung capacity as well as a better tolerance for low air pressure and minimal oxygen.

For radiation oncology technicians or workers operating near sources of radiation, there may be genetic solutions to the risk they are taking. A variant of the NOS3 gene has anti-inflammatory effects in radiation-induced pneumonitis and provides protection against radiation damage. Increased melanin may offer protection against radiation and genomic instability as well as against muscle and bone deterioration. In addition, the NOS3 gene can reduce an astronaut's risk of developing cancer. Moreover, the DSUP and TP53 genes provide increased skin density, protect the DNA and tissues from radiation, and reduce the risk of gene-expression dysregulation, persistent telomere loss, and critical chromosome shortening.

If you happen to be an astronaut anticipating a long mission, some exemplars for gene modifications can serve as a guide for future interventions. For example, a gene called LRP5 is known to increase height and bone density and to protect bones from radiation and microgravity-caused deterioration. A striking example of the power of the LRP5 gene is given in the movie *Unbreakable* (2000), about a man who can survive almost any crash or fight and never gets a broken bone. Similarly, the MSTN gene leads to increased muscle growth, protects against muscular deterioration, and supports a strong and lean muscle structure. When the same version of MSTN is present in dogs, they look like WWE wrestlers.

For astronauts, there are plenty of other ways to improve the safety of the mission. For example, their immune system could be optimized and tweaked to develop faster and with greater flexibility. A well-regulated immune system can adapt to highly sanitized, low microbial-diversity environments like the space station, protect against postflight-induced stress and inflammation, improve intravascular fluid management, and potentially prevent long-term cardiac problems (figure 9.2).

For athletes who need extra endurance or are aging, there are also plenty of options. Instead of inserting genes or components of genes into players, they can get SIRT1 or NAD "gene boosters," which use epigenetic methods to activate these genes for a short period to mimic exercise and provide the body with endurance and better blood flow. In addition, SIRT1 and NAD can help protect the brain and the body from cosmic and ultraviolet radiation. If cells need to be regenerated, the short-term cyclic expression of four genes, OCT4, SOX2, KLF4, and MYC, together known as OSKM, shows signs of partially reprogramming mature cells and reversing aging, maintaining muscular atrophy, and providing greater longevity. The expression of related genes such as UCP2 and 5'-AMP can provide increased cryoprotection (that is, protection against cold), which might be useful for recovery from and/or enabling deep hibernation.

Reprogramming for a New Job

Of course, the great mass of humanity does not consist of astronauts and pro athletes. They're people who hold down jobs that are relatively complex, require a range of subtle skills, and often generate problems of considerable ambiguity.

Although exciting, these "genetic therapeutics" are still only biomedical predictions based on thousands of genomes and samples, not millions or billions. To truly know the efficacy of such genetic predictions for traits and genetic tweaks, longitudinal medical profiling of patients and integration with clinical metadata would be needed to ascertain if such single-gene changes could really lead to persistent and functional alterations in people across a range of jobs. It might turn out that some of these genes or variants have unexpected negative impacts later in life or affect health in ways that could not be predicted from a small set of patients.

Yet genetic fate does not need to be permanent. Recent variations of CRISPR gene-editing technology have demonstrated that epigenetic enzymes can be combined to allow for transient activation or suppression of specific genes. For example, some genetic tools have been created that exploit the power of hybrid human and bacterial proteins, which can

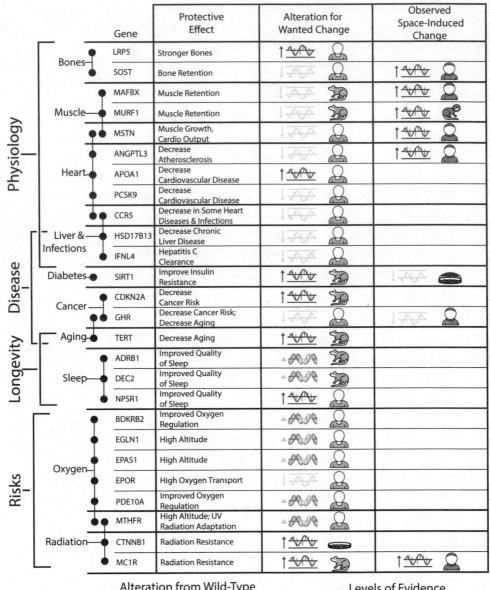

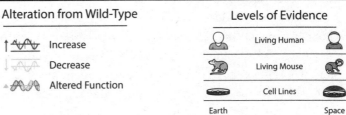

temporarily toggle on genes related to DNA repair and then toggle them off. Studies in 2016–2018 by the Whitehead Institute's Rudolf Jaenisch and Duke University's Charles Gersbach showed that both histones, which provide the scaffolding for DNA, and DNA itself can be reprogrammed to alter gene activity and context. The genome, in short, is a kind of plastic playground.

In 2018, the Defense Advanced Research Projects Agency (DARPA) funded grants for a project called PReemptive Expression of Protective Alleles and Response Elements (PREPARE). DARPA wanted to prevent acute radiation syndrome, more commonly known as radiation sickness, by using epigenetic-editing methods to activate genes before exposure to radiation. This application could help astronauts on long-duration missions; soldiers deployed into irradiated areas, such as after a nuclear explosion; and cancer patients enduring radiotherapy. These studies are ongoing in a variety of labs, including Chris's lab at Weill Cornell and the Innovative Genomics Institute, a joint effort of the University of California, Berkeley, and the University of California, San Francisco.

Given these advances, if all of the genes in a cell become as easy to turn off and on as light switches, then it is simply a question of having the right "genetic electrician" for specific cellular construction projects. This means that the fatalism of genes will eventually be replaced with the transience of genetic and epigenetic editing methods. Any person could be modified to be the best for any job, and it would be harder to predict what someone should or could do (as in *Gattaca*) because that person's genome might be completely different at one point from what it once was.

If you can reprogram nature to fundamentally change it, does this mean that nature is no longer a factor and that nurture reigns supreme? Not

---

Figure 9.2
Genetic improvements to human cells for spaceflight. Experimental data from altered genes and gain-of-function or loss-of-function tests on human and mouse models, both on Earth and on the International Space Station, have shown potential avenues of new therapeutics or protective effects that could improve cardiac function, bone density and function, muscle strength and persistence, and other areas of health. Evidence from living humans, mice, and cell lines is shown for each gene, and ideal alteration types are shown for an increase (dark gray), decrease (light gray), or completely new function (gray).

entirely, but it does mean that nurture (for example, gene therapies) can change what nature means and that nature is (more than ever) malleable through cell- and genome-editing methods. Even though these genome- and epigenome-editing methods and technologies are very new, they reveal the fallacies of previous metrics of performance, employment, and abilities, which were assumed to be fixed at birth. They represent a possibility for newfound "cellular liberty" to enable a person to reprogram any cell to add traits of any other cell, which in theory could enable new job opportunities not previously possible for an individual.

Yet knowing about these genes does not mean that we can engineer all results. Gene therapies today work mostly to alleviate the effects of lethal mutations, and they are advancing rapidly. So what are the risks from what might be called "predictive genetic engineering," a kind of cosmetic sur- gery for the genome? One issue is the same problem that afflicts many AI- enabled assessments: without truly knowing what constitutes a successful employee or student, prediction is a moving target. The risk of a big mistake remains large. Assessment may well involve a cloud of traits, some more predictive than others and all prone to change, fed by many more sources of data than we have today, all the while skirting privacy and discrimina- tion issues. Like medicine, assessment is always personal, which may run contrary to the institutional desire for low-cost, efficient name-gathering and assessment guidelines. But the answer to the assessment question will almost certainly involve more algorithms and more data, not less.

Opening the door to predictive genetic engineering also ushers in the issue of behavioral responses. Many sports have struggled to deal with dop- ing as athletes eager for a competitive edge attempt to artificially bolster their natural attributes or extend them, despite quite clear risks. Doping is a way of gaming the system. Meanwhile, given the money involved in sports, there will always be pressure to dig deeper into the genome and to alter it to improve performance or to tweak the epigenome. That pressure comes from both athletes and management. And as sports go, so, too, does the broader public.

Looming above these questions is the specter of eugenics, which, as we've seen, was viewed for many decades as scientifically rigorous and socially and ethically responsible. Eugenics, of course, was based on a deterministic calculus of heritability (without any knowledge of the gene); nurture was rejected as sentimental twaddle arrayed against the Darwinian realities of nature; and failure could be eliminated by breeding. Eugenics, in the nineteenth and the early twentieth centuries, was convinced that it had answered the assessment question: What are the secrets to success? The answer was, first, the hierarchy of "races" and, second, the hierarchy of "class." Race was static and determinative, not plastic; certain races were superior to others, and even within races, genetics determined the success or failure of individuals. These assumptions were not only scientifically incorrect but sociologically and generationally harmful. And given the tools of today, genetic fatalism is not only unwise, it is simply untrue.

We now know extraordinarily more about genetics, but as the rapid progress of the past few decades suggests, an enormous amount remains to be discovered, one prediction at a time.

10   The Plague of Polling

Short of economic or weather forecasting, the best-known—and most controversial—predictive enterprise is political polling. In the modern sense, polling has been around since the 1930s, and it has grown into a sizable and diverse industry. Over that time, it has accumulated a large, rabid following, particularly during presidential elections, as well as a long list of critics; polls are both cited constantly and attacked continually for their failures. As W. Joseph Campbell elucidates in entertaining detail in *Lost in a Gallup: Polling Failure in U.S. Presidential Elections* (2020), some of the most respectable polling organizations in the United States miscalled presidential races in 1936 (Franklin Roosevelt won in a landslide over Alf Landon), 1948 ("Dewey Defeats Truman"), 1952 (Dwight Eisenhower wiped out Adlai Stevenson), 1980 (Ronald Reagan won in a landslide over Jimmy Carter), 2000 (Bush versus Gore), and 2016 (Donald Trump upset Hillary Clinton). In 2020, the year Campbell's book was published, Trump lost to Joe Biden and claimed massive fraud, once again stirring controversy over polling and pollsters.

In fact, a panel assembled by the main US trade group for pollsters, the American Association for Public Opinion Research (AAPOR), described the polling for the 2020 presidential election as the most inaccurate in 40 years. The 2020 polls overstated Biden's support by 3.9 percent, well above the 1.9 percent error in 2016, and the worst since the 6 percent whiff at Reagan's victory in 1980.

For all the talk of yet another failure, pollsters did get the top line correct in 2020: Biden won 306 of 538 electoral votes and received 7 million more votes than Trump. But that outcome was obscured by the fact that the votes in a few battleground states that went to Biden were far closer than pollsters had suggested, and Trump won a higher than anticipated proportion of votes from some minority groups, such as Latinos. The large number of mail-in ballots whipsawed both sides, and the pandemic roiled the outcome.

Despite the litany of presidential polling failures described by Campbell, the public's appetite for polling has only grown. Polling provides satisfaction for those consumed by the horse-race aspect of politics. Cable news dwells on polls years before actual elections. Cognoscenti famously return again and again to sites that run multiple polls through complex models, such as Nate Silver's FiveThirtyEight, which shrewdly packages results into easily digested percentages, or to poll aggregators that average major polls, such as RealClearPolitics. Another kind of political junkie consults the betting markets, which track the changing odds of a candidate's potential win or loss. Betting markets, like modeling operators or aggregators, feed off polling data; they do not run polls. All this surveying, sampling, number crunching, and projecting is amplified by the media, which creates complex repercussions, feedback loops of excitement or despair (depending on the poll numbers), and a tendency to congeal into a version of the conventional wisdom. Donors queue up or exit depending on polls, and campaigns make key tactical decisions—where to put resources or take them away—that hinge on polling data.

Since 1936, when George Gallup's pioneering poll beat a crude survey by a popular (if fading) magazine, the *Literary Digest*, pollsters have claimed that polling is an objective and scientific endeavor, an amalgam of probability and statistics and mathematical models. (Gallup pioneered more than just techniques: he was quick to publicize his polls by syndicating them in newspapers, and he moved into polling public-policy issues.) Polls are almost never straightforward reflections of the electorate's preferences, however. Because they sample small slices of larger populations, they require a number of assumptions, each of which may skew results one way or the other. Despite the criticism polls receive—they have been attacked from the very beginning—pollsters insist that their increasingly sophisticated statistical-modeling techniques can predict the behavior of fluid, complex electorates, resolving key questions such as who will vote and who will not; how large the ranks of undecideds will be; and whether potential voters are telling the truth about their preferences or will change their minds. (The so-called shy Trump voter, who supposedly resisted publicly admitting to that preference, appears to have been something of a myth.) These conclusions can be presented in many ways. What does it really mean that Donald Trump had only a 2 to 11 percent chance of winning in 2016, depending on the pollster? What does it mean that telephone survey work requires more and more calls as response rates fall? What effect does a polarized society have on public opinion when millions of voters occupy the echo chambers of right-wing or left-wing cable TV, talk radio, and social media?

In fact, traditional polling probably does a better job than many think, at least within a 4 to 5 percent margin for error. Nate Silver, who made a name by picking Obama in 2012—driving the Gallup organization, which picked Republican challenger Mitt Romney, out of presidential polling—has long sparred with those who claim polls are wildly off, particularly given the complexities of the US Electoral College system, which involves 50 elections, not just one. Although Silver is probably right, from a historical perspective there's little evidence that polling has grown significantly *more* accurate over time. Polling today runs complex models through powerful computers, but despite some innovations, particularly in techniques for sampling and poll aggregation, the basic set of ideas driving opinion surveys has remained much the same.

Over time, survey work has shifted from relatively expensive and slow in-person interviewing to inexpensive and faster telephone (or internet and mail) surveys. Telephone polling in particular has seen a dramatic decline in response rates for a variety of much-speculated-upon reasons: resistance to aggressive telemarketing, the spread of answering machines and caller ID, and especially the replacement of traditional landline phones, which have mostly publicly available numbers, with cellphones, whose numbers are far more difficult to access. The Pew Research Center reported that phone response rates dropped from 37 percent in 1997 to 6 percent in 2018, and that was for general polls, which are less upsetting to people than political polls.

Other experts blame an increasingly polarized US electorate for the response problem. For some time, conservatives have appeared to harbor a greater distaste for polling than liberals. Today, that divide may be even wider. The AAPOR panel suggested that much of the error could be attributed to a failure to include voters with less education. The latter were not shy Trump voters; they were instead voters who refused to take polls. Polls, by their very nature, represent established opinion, with polling widely viewed as part of the national "elite" or "liberal" media. Skepticism about "fake news" is routinely aimed at polling, not least by former president Trump, who bashed even the polls taken by his ally Fox News. Even if a pollster managed to get a Trump supporter on a landline phone, that voter would likely have hung up.

In short, traditional polling has been a predictive enterprise for which good data have grown more difficult to accumulate and analyze. Although much of the finger-pointing about presidential polling has a kind of knee-jerk quality, a mounting body of literature generated from within the polling world is asking why polling is not better. Natalie Jackson, who oversaw the *Huffington Post* polling operation that in 2016 projected that Hillary Clinton had a 98 percent chance of winning the presidential election, has argued that too many people think polling is a crystal ball; she believes polls and aggregators need to do a better job, particularly of "eliminating bad or redundant data" and ignoring "low-quality polls." Jackson criticizes both the emphasis on the horse-race aspect of polling—the predictive quality—and the public's failure to recognize that polls are mere snapshots in time. Silver, who dismissed Trump's chances of winning in 2016, has long blamed the media for unrealistic expectations. After the 2020 election, Silver's critique on the FiveThirtyEight blog concluded, "Voters and the media need to recalibrate their expectations around polls—not necessarily because anything has changed, but because those expectations demanded an unrealistic level of precision."

The revealing phrase is "not necessarily because anything has changed." Silver is essentially admitting that in polling not much has changed. And, he says, folks need to get used to it.

The problem with both critiques is that if you eliminate the snapshot (a constant stream of opinion, resembling the betting markets, is economically unrealistic for pollsters) and the horse race (the predictive nature of polling) and succeed in recalibrating the public's expectations, few will care about polls, and the industry will wither.

Data and Uncertainty

Is there a better way?

For more than 80 years, pollsters have attempted to capture mass public opinion by focusing on individuals. Probability and statistics provided the

path toward surveying relatively small and commercially desirable numbers of people who would accurately mirror the larger population. That was
the achievement of Gallup and his colleagues and rivals.

The problem is not the underlying science; the problem is the data. Preferences are tabulated through questions, whether in personal interviews
or on the telephone. As we've seen, response rates have plunged. But how
trustworthy are the data coming from many participants even if you succeed in doing interviews? Do respondents tell the truth? Do they have
firm opinions about candidates or issues? ("Undecided" is a big problem
for predictive clarity.) Will they vote? And if they have a tendency not to
vote, will they admit it? Many pollsters still adhere publicly to Gallup's
ideal of the autonomous, civic-minded voter who carefully digests the news
(mostly from that relic the daily newspaper), weighs choices, and makes
rational decisions—in other words, an efficient market in democratic, well-
informed participation. But that's an unrealistic idealization, especially of
the well-informed voter who can keep up on current affairs despite the
bewildering variety of sources.

People produce polling data, not natural forces or machines; they are
individuals negotiating complex, shifting environments and social mores,
not traders seeking profits. An election, like many aspects of life, is a complex decision that has no right or wrong; it's a decision based on what the
economists John Kay and Mervyn King call "radical uncertainty"—that is,
a future in which voters *in the best of all circumstances* lack information
and knowledge of available options and potential consequences. Much
like investing, voting is never fully rational. As Kay and King note, people
don't normally optimize, as economists have claimed since the nineteenth
century; rather, they cope by making incremental and empirical decisions
based on limited information. It's not surprising that people may change
their minds when pollsters come calling. They may fib about what they're
thinking (sometimes without realizing it). They can be led astray by deceptive questions, particularly those that tap into their biases. They are often
not deeply conversant with public affairs—the proverbial low-information
voters—and may not even know who's running for office.

In 1922, the columnist and public intellectual Walter Lippmann published an explosive book titled *Public Opinion* that questioned the very
foundation of democracy: the belief that individuals can rule themselves.
Lippmann was a newspaperman and during World War I a propagandist

for the US government, and he knew how the papers—then the largest source of information—presented stereotypical "pictures of the world" to "the mass of absolutely illiterate, . . . feeble minded, grossly neurotic, undernourished, and frustrated individuals." That criticism was harsh, but Lippmann was both an unabashed elitist and a proud realist. His answer to the problem was technocracy—rule by experts—which he would write about in a later book and which would triumph with the New Deal and a growing administrative state.

His condescension aside, Lippmann got a few things right about public opinion. Sources of information, presented by a fallible media perceiving an elusive reality while trying to shape news that will attract large numbers of people on a daily basis, are often suspect. And news consumers are often distracted or ignorant of public affairs, particularly in a world that's both tightly connected and increasingly fragmented (and here we're not even getting into social media echo chambers and rampant conspiracy theories). Following current political matters may well be low on their list of priorities. Thus, surveying the public's feelings about candidates or pop stars or fashion trends provides data but perhaps not, as Silver admits, precise predictions.

In the run-up to the 2020 election, Jackson, still pondering her errors in predicting the 2016 result, wrote an essay for *Sabato's Crystal Ball*, an online publication of the University of Virginia's Center for Politics, that began: "Humans generally do not like uncertainty. We like to think we can predict the future." In two sentences, she manages to lash together the two themes that are the focus of this book: prediction and risk, which she (like Frank Knight, John Maynard Keynes, and Kay and King) defines as uncertainty. She admits that the failure of her polling in 2016 led her to consider "what went wrong and the impact of my work." Her conclusion involved two entwined issues. "The first is that we do not really know how to measure all the different sources of uncertainty in any given poll," she writes. "That's particularly true of election polls that are trying to survey a population—the voters in a future election—that does not yet exist." The second involves "marketing those attempts to solve it [uncertainty] to the general public in the form of a seemingly simple probability or odds statement that the public lacks the tools and context to appropriately interpret."

Jackson runs through stubborn uncertainties that plague the polling process, from changes in telephones to suspect internet polls to the ability to identify likely voters. The past, she argues, has ceased to be a reliable guide: "Because both technology and populations are constantly changing, we might not be able to count on error rates in past elections being indicative of error rates in future elections." Thus, she believes that calculating an accurate measure of uncertainty or risk is growing more, not less, difficult.

Her final point, however, may be even more important. Efforts to calculate uncertainty and error rates have made the models increasingly complex. The public, Jackson notes, doesn't understand probability, and the models are "thoroughly inaccessible for those not trained in statistics, no matter how hard writers try to explain it. I tried, too, but my guess is that most people probably closed the page when they got to the word 'Bayesian.'"

What can be done? A vast amount of data about individuals or groups never finds its way into the models of opinion polling. These data are what we referred to earlier as "proxy data"—secondary data sources that may provide clues to what someone might or might not do. Proxy data are secondary data; they are usable if they show a correlation with key issues, such as rising cellphone usage or sudden movements of people. It's like the attempt to anticipate consumer behavior during the pandemic—pantry loading, a rising demand for delivery of goods, nuanced shifts in consumer purchases—that we described in chapter 5. No proxy data can definitively predict the behavior of any individual at any time. But taken in aggregate, combined with other data sources, and run through machine-learning programs, proxy data should, in theory, generate better predictions.

What kind of data are we talking about? Most of it is still being discovered. We have only scratched the surface of, say, the relation between consumer choices and behavior: home location, credit rating, car make and model, education level, church membership, country club attendance, or gun ownership? Then there's demographic, census, and financial data. NLP of social media activity may offer clues to preferences buried among Facebook likes and Twitter posts. Traditional polling will continue to be a useful if not determinative component, but there is an expanding universe of unexplored data sources out there.

A few attempts have been made in this direction, mostly from data science firms rather than traditional polling organizations. With the rise of internet fund-raising, political parties and campaigns have built deep data sets of donors (often small-dollar contributors) and supporters; Trump rallies were opportunities not just to energize his base but to sell merchandise and gather data. That has led to the ability to focus on a more granular analysis of the electorate, not just on national, state, or county preferences but on towns, neighborhoods, even households and individuals. Campaigns use some of this type of granular analysis to marshal their troops, the door knockers and postcard writers, and to spend their money and time more effectively, particularly on media.

And there's an additional factor. The line between knowing what someone may do and getting them to do it has been blurring. The same data that can produce an accurate survey of preferences can be used in that political dark art par excellence: messaging.

This is an important point that applies to many predictive enterprises. The ability to target and understand voters' deeper motives and patterns can be used not only to predict their electoral choices but also potentially to manipulate them. Data, in particular proxy data, can be weaponized. The data may open the door to self-fulfilling prophecies, which may in turn produce paradoxical effects, particularly in nations that claim to be democratic.

Polling as Persuasion

During the 2016 US presidential race, the Trump campaign, which started with a relatively rudimentary campaign structure, made a giant leap into predictive messaging. Democrats had pioneered the use of greater voter data in the 1990s, mostly for fund-raising and targeting. But in 2015 the Trump campaign signed up a UK-based data firm called Cambridge Analytica (CA), which marketed more ambitious plans. CA was backed by Robert Mercer, one of the cofounders of the quantitative investment firm Renaissance Technologies. In late 2013, Steve Bannon, the publisher and filmmaker who would soon become Donald Trump's campaign manager, persuaded Mercer to provide $15 million to start a political strategy firm, which would eventually be formed around a team from the UK military contractor SCL Group. The SCL Group was engaged in selling communications and information-warfare services to overseas clients, mostly in the military and intelligence. The company began working on elections overseas and then on elections in the United States and the United Kingdom. The relationship between SCL and CA was complex, in part to allow CA to operate in the United States but keep Mercer's ownership quiet. Mercer owned 90 percent of CA and 10 percent of SCL.

CA plunged into the project of trying to understand American society by analyzing how cultural and psychological tendencies shape larger ideas, such as politics. In his book on CA, Christopher Wylie, one of the early data scientists at the firm and later a whistleblower on its efforts, described what Mercer was after: "To put it crudely, if we could copy everyone's data profiles and replicate society in a computer—like the [video]game *The Sims* but with real people's data—we could simulate and forecast what would happen in society and the market. This seemed to be Mercer's goal. If we created this artificial society, we thought we would be on the threshold of creating one of the most powerful market intelligence tools in the world. We would be venturing into a new field—cultural finance and trend forecasting for hedge funds." The big dream was known as "simulating society *in silico*"—capturing every individual in a society in a data set on a computer chip.

As Wylie discovered, Mercer was less interested in CA's near-term commercial possibilities than in understanding the US electorate well enough to win elections instead of just predicting them. "Mercer looked at winning elections as a social-engineering problem," writes Wylie.

To achieve this goal required lots of data. In 2015, CA managed to find that treasure trove at Facebook, which had some 3 billion users around the world. The social media giant was a vast repository of personal data, most of it volunteered by its own users. Facebook routinely allowed outsiders access to that data for everything from targeting advertising to allowing outsiders to help the company better understand its own users. As a result, CA was able to download Facebook files on some 87 million US users—a fact that, once it became public knowledge, proved to be intensely embarrassing to Facebook, which initially denied that the data had even been lifted.

CA was not engaged in traditional polling, though its big US election project began with a 120-question personality survey in the form of a quiz, which paid a few dollars to those who filled it out. The survey was not trying to guess public opinion; it wanted to go deeper. The survey, known as Ocean, broke participants down along five axes: openness to experience, conscientiousness, extraversion, agreeableness, and neuroticism. To get paid, participants had to log onto Facebook and download an app developed by a Soviet-born Cambridge University academic, Aleksandr Kogan, who had done academic research on personality profiling. Immediately, the app downloaded the user's responses to the survey quiz onto one table, the user profile onto another, and all the information on the user's friends onto a third. The data from a few hundred thousand participants were then used to build a 100-million-feature data set ensembled by a variety of machine-learning programs; as we saw in chapter 3, ensembling allows machine-learning algorithms to work through a variety of hypotheses, iteratively seeking the most predictive one for the data. One of the architects of that process told the *Guardian* in mid-2018 that the ensembling allowed the group to build 253 algorithms, "which meant there were 253 predictions per profiled record." Those predictions involved guesses about political affiliations, personalities, and lifestyle choices. All of this was perfectly legal.

Cambridge Analytica then married those profiles to publicly available voting records in 11 states, which it would use to refine messaging to individuals on social media. This project was not about prediction but rather about control. Bannon's candidate won.

How effective was CA's Facebook-driven simulation *in silico*? We don't fully know, even now. What we do know is that Trump won by the narrowest of margins in a handful of swing states, a victory that resulted from myriad factors. CA employees boasted that their tools "helped Trump clinch the election" in 2016, and they were caught bragging about it again in 2018, when it was still in CA's self-interest to say that. Headlines in the United States and the United Kingdom repeated the clincher claim while also endlessly discussing Russian hacking, former FBI chief James Comey's late intervention in the campaign, and other factors. In a narrow win, a clincher might be tiny. Wylie himself does not mince words in describing what happened: he insists CA used the Facebook data on voters to create a "psychological warfare mindfuck tool."

Wylie and Kogan believed they were pioneers of a new science of "behavioral simulation" that could "bypass individuals' cognitive defenses by appealing directly to their emotions, using increasingly segmented and subgrouped personality type designation and precisely targeted messaging based on those designations." Beyond these bold claims, however, no one really knows how effective these techniques were. CA was not engaged in a randomized clinical trial; there was no ability to test counterfactuals, such as what would have happened if CA hadn't appeared, or to engage in a direct comparison, as in the NASA identical twins study Chris worked on. An aide in Senator Ted Cruz's campaign, which briefly hired CA in the 2015 primaries, told the news and opinion website Gizmodo in 2018 that one of CA's political products, a voter-analysis package called Ripon that was designed to help classify voters by personality types, never actually worked and was essentially "vaporware." How illuminating was the Ocean survey? As the *Guardian* noted on May 6, 2018, "While this was undoubtedly a highly sophisticated targeting machine, questions remain about Cambridge Analytica's psychometric model. . . . When Kogan gave evidence to parliament in April [2018], he suggested that it was barely better than chance at applying the right Ocean scores to individuals." In testimony before Congress, quoted in the *Guardian* on June 19, 2018, Kogan was even more definitive: "This is science fiction. The data is entirely ineffective."

Even if the specific impact is debatable or very small, there are enough data to show that such techniques can be predictive and powerful enough to sway people's views. A study by Chris Sumner and colleagues in July 2018 at the UK-based Online Privacy Foundation showed that they could not only find subjects for the study based on Facebook's understanding of their social profiles but also predict their tendencies, including their psychological and political leanings. Specifically, they tested whether people could be swayed to be more authoritarian in their views (based on comments, likes, and links shared) and support mass surveillance. They said such psychographic targeting is a "weaponized, artificially intelligent propaganda machine" that is effective because "you don't need to move people's political dials by much to influence an election, just a couple of percentage points to the left or right." In their view, this kind of targeting was a clear return on marketing investment.

Facebook itself has claimed that such targeting works. In a pioneering set of experiments in 2014, known as mood-control experiments, Facebook purposely manipulated user news feeds, which show selected content and updates from friends. The company worked with sociologists to alter the feeds so they would be predominately positive or negative in terms of content. They then observed whether the original recipients shared these stories, photos, and emotional states with friends, who would then "acquire" the mood and spread it further. In the paper published by the *Proceedings of the National Academy of Sciences* about the study, the researchers explained that they wanted to find if "emotional states can be transferred to others via emotional contagion, leading people to experience the same emotions without their awareness." Indeed, people's views could be shaped by these methods, the researchers concluded.

"The data provide, to our knowledge, some of the first experimental evidence to support the controversial claims that emotions can spread throughout a network," the authors wrote. Even though they framed their work by saying that "the effect sizes from the manipulations are small," they were clear about the possible implications, arguing that "given the massive scale of social networks such as Facebook, even small effects can have large aggregated consequences" across billions of users. They even speculated about possible instances of large-scale changes affecting the physiological health of entire populations, noting that "the well-documented connection between emotions and physical well-being suggests the importance of these findings for public health."

None of this academic verification of the effect of CA's methods did the company any good. Cambridge Analytica, which also worked for the winning Brexit "leave" campaign in 2016, was eventually shuttered, brought down not only by charges that it stole the Facebook data but also by recordings of its CEO, Alexander Nix, boasting about using a variety of less sophisticated but traditional means, including prostitutes, honey traps, and bribery, to influence more than 200 elections overseas. But Cambridge Analytica left its mark. Its use of behavioral research and psychometrics remains controversial and, like early cases of genetic testing and the use of autonomous military drones, the subject of considerable debate over big tech's policies on data, privacy, surveillance, and how far predictive technologies can be pushed.

An Attack on Democracy?

Some critics have long viewed polling as an attack on democracy, potentially poisoning that source of democratic legitimacy, voting. In 1996, a journalist named Daniel S. Greenberg wrote a column in the *Baltimore Sun* that effectively summed up his problem with what he called "the quadrennial plague of presidential election polling." Greenberg was hardly a crank. He was a veteran journalist who had helped transform science reporting at *Science*, the journal of the American Association for the Advancement of Science, and who published the *Science & Government Report*. He knew of polling's prediction failures, but that wasn't really what bothered him about what he characterized as an "infestation of polling moving deeper into the electoral system." Greenberg's critique focused on polling results that "are

easily confused with political reality, producing bandwagon effects, heart-
ening the leaders and disheartening the laggards." Polls can make it seem
as if an election is over long before Election Day, undermining "the historic
role of campaigns . . . to educate the voters about candidates and issues."
Polling encourages candidates to alter their personae or issues based on
"voters' anxieties and fears," leading to governance by polling. Worst of
all, in Greenberg's view, were deceptive push polls, which under the guise
of a conventional poll try to influence voters through deceptive questions,
spreading "political poison." (Push polls were the primitive predecessors
to CA's efforts.) How can citizens protect their rights against this insidious
force? Easily, wrote Greenberg: refuse to reply, or lie. After all, small events
can create large errors, which could bring polling down.

A few years after Greenberg's jeremiad, Kenneth F. Warren, a professional
pollster, spent 317 pages of his book *In Defense of Public Opinion Polling*
(2001) reviewing and refuting the case against the practice. His first chapter
went straight to the problem: "Why Americans Hate Polls." He broke the
reasons down into six capacious buckets: polls are un-American; polls are
illegal, if not unconstitutional; polls are undemocratic; polls invade our
privacy; polls are flawed and inaccurate; and polls are (paradoxically) very
accurate and intimidating.

That was two decades ago, an age before social media, smartphones,
mainstream conspiracy theories, and CA's psychometric techniques. Warren's sunny defense of polling, although comprehensive, showed no appreciation for the darker currents already running through modern American
society. (Many of these currents, such as paranoia and conspiracies, have,
of course, long been part of US history.) In fact, the anxieties provoked
by polling are in their own way predictive. Moreover, many of those fears
emerged in more potent form with the new technologies and techniques.

To be successful, prediction technologies at a certain point run into
questions over privacy, as we've seen in all other predictive applications in
this book. They require data unique to individuals, such as their genome
or (far dicier and less developed) the bubbling contents of their minds
and personalities—what the late nineteenth-century psychologist William James called "the stream of consciousness." Prediction of nature is
the subject of that awe-inspiring endeavor known as modern science. We
want to know what the weather will be, how the pandemic will spread, or
when the earthquake will occur. We may doubt that prediction is possible

or may believe that we, like early proponents of smallpox inoculations, are engaged in a rebellion against God's will and so resist the advice of science. Prediction in humans, however, cuts far more deeply and is far more difficult. To achieve a degree of predictive precision or even to develop a better quantitative sense of uncertainty and risk requires an understanding of human impulses and dynamics—and thus raises the specter of control or manipulation.

Artificial intelligence that can be used for prediction and control has proved to be a boon for authoritarians—the double-edged sword that AI critics such as Elon Musk have warned about. In their obsession with control, authoritarian or totalitarian states are often rigid, lacking the kind of free-flowing data that a liberal democracy and a free-market economy—the liberal political economy that the political economist Francis Fukuyama, in a burst of post–Cold War optimism, once saw as possessing a global manifest destiny for democracy and capitalism (figure 10.1). For all their inefficiencies and redundancies, democracies do produce a rich flow of economic, social, and political signals that provide early warnings of social tensions and feed flexible and creative individuals, from entrepreneurs to artists to scientists.

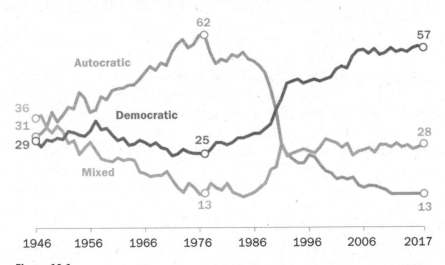

Figure 10.1
Proportion of countries of varying regime types around the world. Data since 1946 show that the proportions of different regimes (autocratic, democratic, and mixed) have fluctuated but that the majority of countries are now democratic. *Source*: Data compiled from Our World in Data, https://ourworldindata.org/democracy.

Fukuyama's prediction, of course, hit some bumps, even if the trend did continue. Liberal democracies no longer seem quite as inevitable as he suggested in 1992 in *The End of History and the Last Man*. Indeed, today we worry about democracies, beginning with elections, which suddenly seem to be fragile vehicles for capturing the popular will. Imagine a set of algorithmic tools, diverse and ample proxy data, and powerful machine-learning programs focused not on manipulation but on learning how to more accurately predict elections. The system would target such key questions as who is likely to vote, how big the undecided pool is, and what deeper psychological factors determine how individuals make decisions. The use of techniques aimed at manipulating voters would be banned. Imagine that over time the failures punctuating the history of scientific polling will fade away, the error rates will shrink, and public confidence will rise. In effect, as predictive capacity grows, the risk of a prediction failing will steadily decline until it approaches zero.

Would that be good or bad from a democratic perspective, meaning not that the "best" candidate necessarily will win an election but that the polls will accurately mirror the sentiments of voters? How would potential voters react to a deep belief that the preelection polls are correct? Except in elections that look extremely close, why would they bother to ponder public issues or vote, except as a sort of civic gesture or a comforting rite? (Today, there's an analogous situation in markets, where increasing numbers of investors opt to buy indices without applying any effort at research or analysis.) This is a complaint long made about conventional polls—that they can make or break candidates needlessly or, more acutely, that calling elections can deter people from voting in states whose booths are still open. If the polls become extremely accurate, will millions simply not bother to vote, believing that the polls aren't wrong? Falling voter participation tends to introduce volatility in the results, like a stock with a small float of shares or like primaries or runoff elections. And what about governing? If prediction gets so precise, why not govern by poll, going directly to the people and getting rid of the leeway traditionally afforded to elected lawmakers to make decisions in a republic governed by representation?

Those questions take us to a very different world, a long way from the one the US Founding Fathers envisioned—in fact, to a democratic reality that they (and later Lippmann) feared. Politics and governance are enterprises engaged in coping with an uncertain future; polls are flickering flashlights in the dark. Lippmann was essentially right about a democratic citizenry uninformed on many important matters, in particular economics, science, and foreign policy. But he may have misjudged the potency of his solution, which was to find experts to tackle issues that in some cases might have no clear-cut solutions, that run roughshod over popular conceptions of fair play or morality, or that require sacrifices by voters. (Think of the difficulties of doing anything about a relatively straightforward prediction problem such as climate change.) In a democracy, politics is ambivalent about prediction: on the one hand worshipping market sages or political commentators who wear the mantle of prescience (until they're wrong enough times) but on the other resisting limitations on free will and incursions into the individual's autonomy. Prediction that eliminates risk and uncertainty may require the kind of personal-data gathering that can feel like a transgression (and in a few cases already requires payment). Moreover, the line between prediction and control—no access to data, no insurance—is often a contested one.

All this raises fewer questions about whether enhanced prediction is possible than about the effects of the backlash to it. There is no doubt that improving prediction promises enormous benefits in any number of areas, shrinking risks that have hung over humanity since prehistory. But it also brings with it new problems and risks, which we have highlighted throughout this book. The drive toward better prediction clearly increases the appetite for more and better data, leading to recent critiques such as *Age of Surveillance Capitalism* (2019) by Harvard Business School's Shoshana Zuboff, who argued in a *New York Times* commentary in 2021 that an "epistemic coup" has been perpetrated by the big tech companies, particularly with respect to the kind of data that drive many advanced prediction technologies. Zuboff believes that if democracy is going to survive, we must regain control over our personal data—"over the right to know our lives." Her solution to the "coup" is for democracies to take back commercial control of data and resist the encroachments of technological surveillance, much as Daniel Greenberg advised folks weary of pollsters telling them what to think not to respond to them, or to lie. Zuboff's somewhat apocalyptic scenario is an illustration of the kind of feedback loops that can be set up by transformations as profound as the enhanced power of prediction.

11   Free Will, AI Jobs, and the Ultimate Paradox

The ancient Greek myth of Prometheus is part family drama, part political allegory, part soap opera. Prometheus was the Titan god of fire and the nephew of the Titan king Kronos. The Titans favored mankind, who lived in a golden age without sickness, pain, work—and women. Zeus, the son of Kronos, eventually turned on his father, forcing Prometheus to choose between them. Prometheus backed Zeus. With Kronos banished, however, tensions mounted between the two gods. Prometheus, a patron of mankind, tricked Zeus, who was more skeptical of men, by allowing them to keep the best part of animals used in sacrificial rites. Zeus, angry, took back from men a key technology associated with Prometheus—fire. Prometheus promptly stole back the fire, hidden in, of all things, a hollow fennel bulb, and returned it to mankind. Zeus escalated, creating the first woman, Pandora, with her box of mischief and unhappiness, and made mankind go to work.

Zeus then ordered Hephaestus, the god of blacksmithing, to chain Prometheus to a rock at the far edge of the world. Every day, an eagle arrived to eat his liver, thought to be the source of emotion, but every night Prometheus's liver grew back, only to be consumed again the next day. The half-human son of Zeus, Heracles, eventually killed the eagle and freed Prometheus after negotiating with Zeus; fire has remained in the human world ever since.

Like all of Greek mythology, the Prometheus story and its manifold meanings emerge from works by men: Homer, Hesiod, Aeschylus, even Plato. In their hands, the myth became a commentary on rebellion and authority, crime and punishment, hubris (or excessive cleverness) and tragedy. But in romantic literature many centuries later, Prometheus also became a symbol of man's refusal to succumb to the larger order of things and of the power of science and technology, which seeks to dislodge divine power. Prometheus became more akin to a cosmic disrupter than to a divine trickster—a human cry against divine limits. Mary Shelley's subtitle for her novel *Frankenstein* (1818) was, aptly, *The Modern Prometheus*, referring to the young scientist Victor Frankenstein, not to his Faustian creation.

For the Greeks, prediction was a divine gift. The name "Prometheus" is usually translated as "forethought," though this god does not appear to have foreseen his own destiny or mankind's. Even Zeus, who had access to portents, lacked absolute clarity about the future. Part of his problem with Prometheus was the latter's refusal to clue him in on a secret about his own fate, which, in the end, involved women. What the Prometheus story illustrates is the complexity of prediction and risk as well as of free will and determinism. The immortal gods—even Zeus—appeared to act with free will and agency, yet they understood that there were limits to what they could know and how far they could go. Zeus would not rule forever. Uncertainty persisted even on Olympus.

Thinkers have been debating free will and determinism for almost as long as human foresight has existed. At the heart of the debate are key questions that resist black-and-white answers: What does it mean to be human? How do we truly know we are acting freely—that we aren't really machines? From those questions emerge religion, philosophy, and a bewildering array of perspectives, opinions, and beliefs. Today, as we enter the Age of Prediction, those questions recur, particularly around the meaning and possibilities of artificial intelligence.

A World of Certainty

For ten chapters, we have been dancing around this subject. In this book, we set out to describe the near-term development of the Quantasaurus—looking ahead, predicting 20 or 30 or 50 years into the future. We have described a world where prediction grows more accurate in almost all domains. It's a world of autonomous robots (cars, drones, vacuum cleaners) and precision medicine. It's a world of algorithms—billions and billions of them, each in some sense embodying a prediction—and of relentlessly rising levels of data. It's a world that features the possibility of ever more intimate monitoring of people and things, often in real time. This is a world that includes a deepening understanding of the genome, the epigenome,

and the microbiome, each of which opens pathways to more accurate diagnosis and more effective personalized health treatment. It's a world where
the urge to "improve" the human stock will grow stronger and where the
ability to simulate human beings in silicon (*hello, Alexa!*) and blur the lines
between human and machine grows apace.

On the other side of the ledger, certain kinds of risk will recede, and in a
generation they may well be forgotten, for good or for ill. Who today wor-
ries about bubonic plague? Not even smallpox inspires the dread once felt
by all parents. Vaccines for new diseases can be developed in a matter of
months instead of in years or decades. To a certain extent, we will come to
trust the opaque logic of the machine, much as we accept the semiconductors at the heart of our computers and our phones, the instructions rattled
off by our GPS system of choice, air travel, vaccinations, and robotic surgery
even on organs as delicate as the brain and the heart.

We are not describing a utopia, but rather a reality. Advances in prediction bring good and bad. As some risks shrink, others will grow: threats to
privacy, economic disruption, epistemic confusion, antiscience sentiments,
mass surveillance, authoritarianism. There is no evidence that risk overall
is about to evaporate. In fact, there are suggestions that, given the role of
Homo sapiens at the center of the prediction-and-risk equation, the reduction of some risks spawns other, different risks and in some cases more
dangerous, more complex risks—*unknown* risks, even existential risks (for
example, hyperintelligent AI). These new risks are especially difficult to
handle when they are things we have never confronted, such as more novel
pathogens or autonomous swarms of missile-laden drones. The speed of
change accelerates, which is a class of risk all by itself.

What happens to risk when prediction approaches absolute certainty?
For most predictions, this absolute state of accuracy probably will *not* occur
in 30, 50, or 100 years, if ever, but that's part of the deceptively strange
relation between prediction and risk that we have been exploring. This is
the Bayesian confidence we all feel, instilled by experience, that flipping
the switch will always turn on the light as we enter a room. What happens
when the same overwhelming certainty appears in medicine, employment,
or warfare?

We can only speculate, not *know*. Igor, for instance, believes that such a
situation may well create strange effects that could be profoundly unpredictable. He argues that you can never eliminate all risk or uncertainty. In

Igor's field of quantitative investing, absolute prediction would eliminate risk, and asset values would simply move up in straight lines like interest-bearing risk-free accounts. The asset prices would go up until the rate of return produced by them equaled the risk-free rate. Essentially reward would be eliminated.

Eliminating reward is a pretty big risk if you're an investor or an entrepreneur. As the economist Frank Knight insisted more than a century ago, the energy source of entrepreneurial dynamism is uncertainty, which is a bit of a paradox but for the world at large a profitable one.

Different Views

This takes us to issues that the two of us do not always see eye to eye on. They're related to the kind of quantum spookiness that seems to accompany absolute prediction. People may create algorithms, but those algorithms have to be processed by powerful computing machines, and as machines learn to learn, they will grow more and more opaque: unpredictable or at least unknowable. We are not suggesting a science-fiction dystopia of machines running amok—at least, not for some time. But predictions will increasingly pour forth from machines in forms we will often not recognize or even think much about, and when we do think about them, we may struggle to untangle their logic. Machines can be as opaque as people, particularly because machines think in real time as a process, continually recalculating and rethinking. Earlier, we discussed how central AI will be to an advancing regime of prediction. Chris embraces the possibility that machines will one day be able to capture all that makes humans human: self-awareness, a moral sense, curiosity, forethought, and stewardship of life. Igor takes a different view: machines are machines; they're not human.

Igor raises the question of what happens if you achieve 100 percent predictive accuracy. That's a state he finds deeply counterintuitive, like some aspects of quantum mechanics. What if machines monopolize all the areas that can be predicted? Such a world would leave humans to deal only with the matters in life that are *not predictable*, which is a scary place to be. What would these matters be? Certain aspects of markets, war, politics. In short, Igor argues, as AI gets more predictable and as we increase the precision of those predictions, humans will grow less predictable. That's a paradox.

What do we mean when we talk about prediction and paradox? We are not referring to ordinary and anticipated consequences, good or bad. For decades, many predicted that key technological advances, from genomics to AI, would create privacy issues for individuals. The question normally devolved to how the trade-off between privacy and data could be managed, legislatively or legally, and that effort continues. A paradox is an unexpected or counterintuitive reaction. One such paradox occurred in the rise of social media. The pioneers of services such as Facebook, Twitter, and TikTok, which link large numbers of individuals online, proclaimed that social media would create communities, enhance sharing, unleash creativity, and promote democracy. Some of those things did occur, but there were other, paradoxical effects, such as cat videos and "deep fake" TikTok videos and metaverses, each with their own realities. Social media provided a stage for self-promotion, a release from the constraints of more conventional methods of social interaction, and a kind of epistemic liberation that led to literal revolutions like the Arab Spring in 2010. Yet they also created echo chambers, spawning disinformation, polarization, and at times a kind of psychological state of nature. They nurtured both democratic uprisings and authoritarian crackdowns. The aggregation of millions of people's activities and emergent communities occurred on the same platform that could deepen and sharpen people's loneliness.

A paradox is, as Igor often says, spooky. When he speaks of "quantum effects," he's referring to the breakdown in determinism that occurred as physicists explored the world at increasingly smaller scales. The natural world that we observe appears to be predominantly deterministic; Isaac Newton thought so, and Albert Einstein, with less certitude, hoped so. The law of large numbers smooths out the random, indeterminate wrinkles. In a deterministic system, one thing always leads to another, and cause and effect can be quantified. Planets circle suns, gravity brings the book you just knocked off the table to the floor at a rate you can calculate, and a baseball hit with enough force leaves the ballpark. We can send a rover to Mars and land it on a metaphorical Martian dime. But quantum effects are strange; particles behave in unusual, even self-contradictory ways. Heisenberg's uncertainty principle begins with two fundamental questions: Why is it so hard to measure both the speed and direction of a particle? When one prediction improves, must another prediction fail?

Let Chance Pour In

Ian Hacking ends his history of the science of probability in the nineteenth century, *The Taming of Chance* (1990), with a profile of a unique character, Charles Sanders Peirce. An American polymath from Cambridge, Massachusetts, Peirce spent years at the US Coast and Geodetic Survey, then taught at Johns Hopkins. He is paired with William James and Oliver Wendell Holmes Jr. in Louis Menand's *The Metaphysical Club* (2001), a book about the development of American pragmatism, probability, and statistics, and came to reject the popular mechanical physics and metaphysics of the age. As Hacking writes, "Peirce denied determinism." He was among the first to use randomization in experimental design and became a leading thinker in metrology, the scientific study of measurement (and, inevitably, of error). He provided a rationale for statistical inference and was a theorist of logic, language, and signs. He believed in "absolute chance," states Hacking. "*He opened his eyes and chance poured in* [Hacking's italics]." But above all, Hacking concludes, "he conceived of a universe that is irremediably stochastic."

What does this mean? Peirce's "answer had many parts, and fitting them together—in a form consistent with his belief in the existence of a personal god—became the burden of his life," Menand writes. "But one part of his answer was that in a universe in which events are uncertain and perception is fallible, knowing cannot be a matter of an individual mind 'mirroring' reality. Each mind reflects differently—even the same mind reflects differently at different moments—and in any case reality doesn't stand still long enough to be accurately mirrored. Peirce's conclusion was that knowledge must therefore be social."

When we talk about the paradoxical aspects of enhanced prediction—and 100 percent is as enhanced as it gets—we are normally talking about the prediction's effect on human behavior, not on photons, electrons, or protons. The human factor enhances the uncertainty. Moreover, embedded within a paradox is a hard kernel of unpredictability: risk. Paradox is less about the prediction itself than about the effect of that prediction on human behavior and the possibility that the effect may undermine or alter the conditions that shaped the prediction in the first place.

Machines can already mirror people. Today, language-translation algorithms such as GPT-3, which can generate human-sounding text, are darlings of AI because they can translate between languages and match mannerisms, dialects, and colloquialisms, making it appear as if the machines grew up in a particular neighborhood—say, South Boston, Massachusetts, or San Antonio, Texas. They are eerily good at imitating speech patterns, can be taught to write poems and computer code, and can even craft sonnets after being fed Shakespeare's works. Researchers and programmers in the field, such as Andrej Karpathy, former senior director of AI at Tesla, first noted these abilities. Karpathy, who builds AI algorithms and recurrent neural network tools, writes blog posts with titles such as "The Unreasonable Effectiveness of Recurrent Neural Networks." He calls such network tools "magical."

The word *magical* suggests something is happening, but we don't know what it is. Let's call it paradoxical. If algorithms and predictions can reach a state where they are correct 99.99999 percent of the time, they approach an asymptote of a nearly 0 percent chance of error. Machines can approach perfection, at least on simple tasks; one can imagine autonomous Uber or Lyft cars always arriving within one or two minutes or less. Margins of error for transportation would eventually be reduced to a matter of seconds. Just such cases have already happened. In 2017, a Tokyo train left the Minami-Nagareyama station 20 seconds early, at 9:44:20 a.m. instead of the scheduled 9:44:40; the next day the train company issued an apology for a "prediction infraction."

It's a straightforward prediction to say that more and more companies will face pressure to be extremely precise in their temporal and logistical predictions. "Just-in-time" production already exists, and it may get better, eliminating some risks and creating others, as we've seen in a gigantic way during the COVID-19 pandemic. The breakdown in supply chains as the world tries to recover from the coronavirus is an unintended consequence of the pandemic. There is a long history to just-in-time production: the coming of the railroads, with their timetables and national route networks, set off not only a communications revolution (beginning with the telegraph) but also a revolution in accurate timekeeping (which necessitated the division of the United States into time zones). This continued a century later with the rise of United Parcel Service, FedEx, and, of course, Amazon.com. The ability to order online with a click provided the gratification of faster delivery. Time frames for the delivery of mail, packages, and other material used to be weeks, days, or the next day (the latter only if you were willing to pay more for the delivery), but we're heading to ever-tighter windows. In 2017, Amazon launched a drone-delivery service called Prime Air with a goal of getting an average "click to delivery" of 15 to 30 minutes. If that happens, a person will be able to order a new outfit online, shave, take a shower, and then don new clothes from a box leaning against the front door.

In 2020, Walmart and Quest Diagnostics rolled out a drone-delivery service for home collection of viral tests for COVID-19, hoping to provide easier, safer, and faster results to those who thought they might be infected. Competitors immediately sprang up—Flirtey and Vault—and it is likely that the drone-delivery race will continue to expand beyond COVID-19. Eventually, any diagnostic test approved for home collection or home use could be quickly delivered to a person's home, including tests for pathogens, pregnancy, sexually transmitted diseases, and chronic conditions such as diabetes and AIDS. Companies are beginning to roll out tests for liquid biopsies, where blood samples can be sequenced to detect a range of mutations or in some cases methylation patterns; these tests not only offer clues to cancer resistance but also provide early warnings. Many of them may end up as home diagnostics.

Given these technologies and the ease of online ordering for tests, the medical relevance of any molecule in your body could be ascertained within minutes, which will fundamentally change how people interact with health care. More of health care will become a service provided at any time and in any place, and some care is already trending that way. Telehealth systems slowly emerged in the late 2010s, but by March 2020, as the pandemic was hitting, they grew 154 percent from the previous year, according to the US Centers for Disease Control and Prevention. This trend will continue, with customers expecting instant access to medicine, testing, and medical information. The risks of medical complications from delays or limited access to service will presumably shrink; with home testing, the risk of passing a contagious pathogen between patients heads toward zero.

As risk falls away, behavior will change. Tolerance for failures of prediction or for "prediction infractions" will decline. People will want predictions to help in all areas of their lives. They will expect medicines to work almost 100 percent of the time and new therapies to quickly arrive for new problems. They will seek meals customized to their own bodies, genomes, and microbiomes. They will want weather to be accurately forecast out to two or three weeks, then months, and within a degree or two. They will demand instant gratification in more areas of life and to be able to see further and further into the future.

This expectation will affect not only companies. Everyone will find they need to perform more predictively and within tighter time windows.

Even if predictions are not 100 percent accurate, they may get close enough to dramatically change the world and how people interact with it. As predictions improve, the elements of life that cannot be well predicted will be left for humans to ponder, as Igor asserts. These elements will be the most elusive, random, and unpredictable aspects of life. And there will almost certainly be resisters and dropouts from the demands of prediction and data, who will add a random element to further predictive endeavors. Some people will resist.

A Question of Human Nature

Absolute prediction can be repressive; the future shrinks to one inevitable option. Absolute prediction produces a deterministic world in which humans, in theory, have no agency, no more free will than a billiard ball, and thus, depending on your perspective, less humanity. Human behavior becomes mechanistic—that is, like a fully determined machine. By definition, the algorithm already knows all the choices you will make. At that point, machines and humans will be, from a philosophical standpoint, very similar.

There's no inevitability to this predictive evolution toward the absolute, however, unless you're what philosophers refer to as a hard determinist or someone, like the seventeenth-century Calvinists, who believes humanity lacks free will and that life is just a very long set of fixed causative effects known by an omniscient God.

Chris is an optimist. He believes that prediction in many areas will get dramatically better, driven by advances in AI, but that humankind can avoid the darkest possibilities. He argues not only that machines can attain level 4 self-awareness and level 5 awareness-of-species self-preservation, but also that machines driven by such sophisticated AI can be limited in many of the same ways as humans. Machines can be designed to include elements of random decision-making that would make them unpredictable, individual, and imperfect—as ornery, irritable, and erratic as humans. Randomness could be built in, as could intuition and emotion. (It may well be that creativity and resourcefulness are products of uncertainty and randomness.) Of course, even if self-aware machines were to become a reality, they would fall into the category of "creatures" that can be modeled and predicted. It is possible that machines might even learn to break rules in a cyber reenactment of Prometheus's story or of the biblical parable of original sin: man, woman, snake, fruit—and out of the garden and off to work you go. This newfound liberty of unpredictability might even allow machines to commit crimes with forethought and (somehow) be held accountable for them. Thinking of that requires some conceptual gymnastics. Remember that, as noted in chapter 8, the US Defense Department refuses to allow drones and bots to fight autonomously because they lack a conscience and moral self-awareness. They cannot feel the pain of transgression. But what if they could?

Although level 4 or 5 AI might possess the appearance of free will—determinists argue that humans possess only the illusion of freedom—such machines may well have advantages over their human creators: they are faster thinking, hardier, smarter. At that point, why keep the flesh-and-blood version around?

Of course, philosophers have never agreed on human nature when it comes to free will and determinism, and there's no sign today of consensus. As a result, there remains a question of how to apply free will to highly determined AI-engineered systems. For now, machines and algorithms remain exemplars of "hard determinism." But when AI algorithms get sophisticated enough to become autonomous, or self-aware, what moral framework should we provide to them? Can we program them to have soft-deterministic states, or will their programming condemn them to exist as code and hard determinists? Or can we program free will? Where do we draw the line between determinism and free will in machines if we still struggle to define that line in humans? Should we fill the machines with tales of Greek myths, like the one about Prometheus, or with the stories of Buddha, Moses, Abraham, and Jesus? And what would doing that mean?

The algorithms might well become so smart that they approach human-level ingenuity, where they can think like humans: writing stories, crafting new concepts, showing emotions such as remorse, procreating, evolving. This future world of emotive and unpredictable AI would disrupt humanity's unique place in the universe as the only beings that we know of who struggle with such epistemic questions. At the same time, humans might gain new genuine friends to meet at the pub to gripe about the universe.

Anticipating Disruption

One expected impact of the Age of Prediction is economic disruption; such disruptions have clear precedent. Indeed, since the first Industrial Revolution was unleashed in England, predictions of mass joblessness, economic dystopia, and mass privation have been common—and some have come true. But not all. The early nineteenth-century Luddites swept across the English countryside smashing the textile looms that threatened them. In the mid-nineteenth century, Karl Marx presented his vision of class warfare that would force industrial workers into poverty and misery and, eventually, to revolution, which would result in the withering of the state and the triumph of the proletariat in a communist paradise of equality and freedom. Not only did such a utopian communist state fail to emerge (though revolutions did), but, some argue, the closest example of a workers' paradise also flickered to life, albeit intermittently, in the United States and postwar Europe.

Of course, it didn't last; paradises never do. In fact, theoreticians of capitalism, from Adam Smith in the eighteenth century to Joseph Schumpeter in the twentieth, predicted that the combination of a free market and technological progress would continually destroy old jobs and create new ones and new wealth. Schumpeter's "creative destruction" has been a rallying cry of capitalism since he coined the phrase in his book *Capitalism, Socialism, and Democracy* in 1942. But Schumpeter also foresaw the seeds of destruction in capitalism's very success.

These tensions exist today, particularly when it comes to AI. Consider the employment destruction that might occur with the widespread commercialization of autonomous driving, which is simply the application of predictive algorithms to vehicles. In the United States alone, there are 3.5 million truck drivers, more than 200,000 taxi drivers, and 750,000 Lyft and Uber drivers; the latter group appeared in the past decade thanks to another technology wonder, the mobile internet, and brutally disrupted the taxi industry. If self-driving cars and trucks were widely deployed, this loss of roughly 4.5 million jobs would represent unemployment for 3 percent of the approximately 150 million active workers in the United States.

Of course, car companies insist that such a future will have a wide range of benefits for former drivers, passengers, and society at large, including less traffic, faster commutes, and safer rides. If there were fewer accidents, fewer cars would need to be repaired and maintained. There would be better air quality and an improved quality of life, especially for city dwellers. Workers might get back some of their commuting time; they might be able to spend more time with their families, leading to lower divorce rates, improved happiness, more sex, and greater well-being for millions of people.

Or maybe not.

The rosy scenario neglects the broad network effects of such a radically altered economy. For example, fewer car repairs mean many auto shops would close (the way electric vehicles have endangered gas stations). The oil and gas industry, already battered by technological and climate change, would suffer. Families might have less need for childcare. Factories could be run by robots, and AI could take over many service tasks and, increasingly, white-collar jobs.

These kinds of broad social predictions are among the most difficult to make, particularly given the sheer number of variables and the unimaginable amount of data required to truly model a society. (Plus, by the time you're done with your model, the world has changed and introduced new risks.) Two inputs are particularly uncertain and protean: technological progress and human behavior. Marx got a lot right, but he never really could take into account a continually shifting technological base and, of course, the frailties of people when it comes to politics and power. He never envisioned the mass starvation precipitated by Stalin in the Soviet Union or nearly anything about China and the rise of its Communist Party, including its move toward capitalism. He did not imagine the genocidal tendencies of some totalitarian states. He also could not have imagined the rapid increase of the purchasing power of citizens in modern democracies (figure 11.1).

That remains the case today. In 1995, the American social theorist Jeremy Rifkin published *The End of Work: The Decline of the Global Labor Force and the Dawn of the Post-market Era*, which contended that worldwide unemployment would increase as information technology eliminated tens of millions of jobs in manufacturing, agriculture, and services. Rifkin predicted that "a small elite of corporate managers and knowledge workers would reap the benefits of the high-tech world economy" but that "the

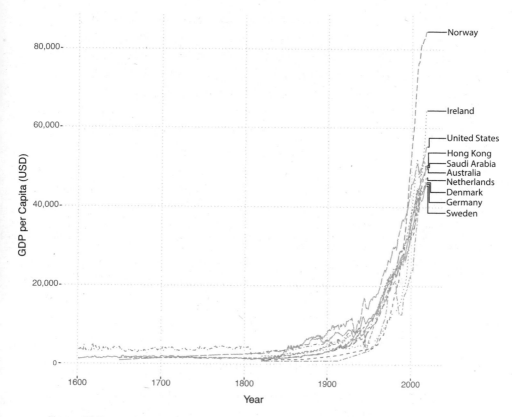

Figure 11.1
Gross domestic product (GDP) per capita, 1600–2016. The GDP for most countries remained flat until the mid-1800s and then exponentially increased. *Source:* Data compiled from the University of Groningen's Maddison Project database, https://www.rug.nl/ggdc/historicaldevelopment/maddison/releases/maddison-project-database-2020?lang=en (CC BY 4.0).

American middle class would continue to shrink and the workplace become ever more stressful."

Rifkin was partially right, mostly about trends that were already clear in 1995, such as automation. Deindustrialization had already begun, hollowing out the industrial Midwest and sending jobs overseas, where labor costs were lower. Inequality had been increasing in the United States since the 1970s, when the postwar boom ended, and the resulting social and economic stresses over the following decades appear to have fueled some of the populism that swept Donald Trump to the presidency. Wages, as measured by purchasing power, stagnated for the middle and lower classes. However, families at the ninetieth percentile have seen their inflation-adjusted wealth increase fivefold, and the wealth of those at the ninety-ninth percentile has grown sevenfold, according to data from the US Treasury Department, the Urban Institute, and the US Bureau of Labor Statistics.

But Rifkin was wrong about worldwide unemployment and a post-market era. Much of the loss of industrial jobs in the United States and Europe came from a liberalized global trading regime, engineered by the United States, that fed new sectors in the developed world: information technology, consumer businesses, and a massive service sector. Postindustrial jobs—jobs held by what Rifkin called "knowledge workers"—boomed, putting a greater emphasis on education. Consumer consumption dominated growth. With the exception of an occasional downturn such as the financial crisis of 2008 and the COVID-19 pandemic of 2020, markets grew larger, more powerful, and more deeply integrated into the global economy than they had ever been. In fact, the single largest factor driving inequality was a global network of markets in financial assets such as stocks, bonds, and commodities. Yet, for all of the stresses, unemployment remained relatively low; in what's now known as the emerging markets, there is a vast and growing middle class around the world—a demographic revolution. This has been particularly the case across Asia, where growth has exploded: in China, Taiwan, Singapore, and South Korea, as well as in Indonesia, Malaysia, and the Philippines. Japan, of course, was already a developed economic power.

Rifkin also predicted that as jobs in developed economies declined, a new, third sector of the workforce would grow built on voluntary and community-based service organizations. He argued that these entities would create new jobs, possibly with government support, to lead the reconstruction of inner cities and drive the distribution of social services. To finance such a society, Rifkin advocated scaling down the military budget, enacting a value-added tax on "nonessential goods and services" that are consumed in particular by the ultrawealthy, and redirecting federal and state funds to provide a "social wage" in lieu of welfare payments to third-sector workers.

A quarter century after the publication of *The End of Work*, there are no real signs that market economies (or military budgets) are in decline, that a third sector tackling inner cities has developed, or that a value-added tax is about to come to the United States, though it's common across Europe.

It is true that a variation on the concept of a social wage, universal basic income (UBI), has surfaced recently as one way to plug the wealth gap. The idea behind a UBI has been around for some time now: guarantee everyone at least a minimum income. It's the idea behind the social welfare states of Europe; the free-market hero, University of Chicago economist, and Nobel Prize winner Milton Friedman once proposed a "negative income tax" for the poor, which would create an income floor. In the 1980s and 1990s, Alaska and Norway created large oil funds that would either directly disburse funds to residents (Alaska) or be used to fund pensions (Norway). Academic studies on UBI experiments have been conducted in Namibia, Finland, Canada, and Spain, with mixed results.

A common argument against the UBI is the paradox that people might not work as much if they aren't motivated by the threat of unemployment—a common argument on issues as varied as minimum wages and pandemic relief. This effect, for all its polemical power, is very hard to quantify, but there may be some truth to the argument: a Canadian UBI study did show slight reductions in hours worked for participants, though the reductions were not spread evenly. In fact, the only two groups that worked significantly less were new mothers and teenagers. A paper that examined data from the University of Manitoba economists Derek Hum and Wayne Simpson showed that new mothers spent this time with their infant children and that working teenagers put significant additional time into their schooling. Overall, the "lost work" was only a small percentage of total work of the previous years and was time arguably spent quite well.

Another critique is that a UBI would stop the natural disruptions and movements of the workforce, which some economists see as an integral feature of capitalism. Schumpeter wrote in *Capital, Socialism, and Democracy* that "capitalist reality is first and foremost a process of change." But there are many sources of change in the modern world, and arguing that capitalism will wither if the poorest members of society aren't forced to work is a stretch. The behavioral forces at work here are turbulent and fluid. You might call them stochastic.

The Final Paradox

Writing a book about the Age of Prediction requires humility about the mysteries of this world—and, more important, some sense of the plasticity of human behavior. People are shape-shifters, like a number of Olympian gods, including Zeus. We often view humankind as chained to a rock, like Prometheus, our fetters forged by evolution, material self-interest, poverty, ill health, and bad luck. In fact, what's most striking about people is their ability to adjust, change, evolve, remember, and forget. Prometheus does not always appear to be a paragon of virtue, but he shares something with humankind—not just the ember of fire in the fennel bulb but also the cleverness, the slipperiness, the trickery. One can imagine he was glib. And he was, of course, associated with that very human trait forethought, which, as we have seen here, makes its own demands on behavioral change. Why, given that forethought, did he play that trick on Zeus? Was it because he felt an impulse to resist the determinism of the all-powerful and create a space for his own free will, even if he could envision the consequences? Or was he passing on that free will to mortal men?

The Age of Prediction has been coming for decades, even centuries, and it is now upon us. Our Quantasaurus is growing up. And we are just beginning to grasp the challenges ahead. We are not gods, but we possess increasingly powerful tools of prediction, unprecedented genetic engineering technologies, and ever-increasing computational power that is impacting almost all facets of life and society.

Here's our final paradox to ponder. Perhaps the most profoundly difficult predictive task in this world is the *prediction of prediction*. How will we know that our predictive capacity is truly predictive or simply the elimination of all other paths to the future? To know, we need to continually test, in the Bayesian sense, all our results and predictions, and closely follow the changing risks. The possibilities are amazing, but so are the risks. The fire of prediction cannot be snuffed out now, nor should it be. The risk of being challenged and charred by our algorithms, while dangerous, is worth the illuminated future.

Afterword: The Future of the Universe

Texas is too cold. California is too hot. The Gulf Coast is too wet. The Southwest is too dry. Greenland is melting; so is Antarctica. A vast amount of data is suggesting that the planet is getting warmer. Carbon dioxide (CO_2) levels and global temperatures are rising, which is worrisome, given that CO_2 is the greenhouse gas that keeps Venus at an average temperature of 837 degrees Fahrenheit, compared with Earth's balmy 59-degree average. Worse yet, the impact of the rise in global temperature lags behind total CO_2 levels, and downstream effects on sea levels are years behind the temperature rise. This means the risks we measure today will not be felt for years to come.

Climate change encompasses much of the relation between prediction and risk that we have explored in this book. It is a physical change caused by human behavior. That's the bad news. That's also the good news: we can understand the physical mechanisms of climate change. (Imagine how terrifying it would be if we had no clue about why the climate was changing or no models.) Although the climate is a planetary-scale system and is obviously complex, we can still model it—and predict it. We can roughly quantify the risk. Humanity can take steps to identify, restrain, even roll back that risk. New technologies can be mobilized; behaviors can change.

But, of course, that's the rub. The consensus on the science of climate change is strong; the public perception is less so. Individuals, groups, and nations have their own self-interests and perceive the risk differently; reactions are subjective, relative, changeable, and variable according to time horizons. Moreover, the issue is not just a matter of diverging self-interests. As with the COVID-19 pandemic and vaccinations, the consensus creates a backlash among those who have come, for whatever reason, to distrust science or (more generally) expertise. As a result, the effort to combat climate change is complex and shifting; the uncertainty is greater than if it were merely a scientific problem.

Many threads make up this giant snarl: political, regulatory, and market responses; technological innovations; varying perceptions about sustainability. But the shift to electric vehicles has begun. There are active projects to capture, modify, or reprocess carbon so that we can control global levels of CO_2 and other gases. The US Department of Energy now supports several Clean Energy Hubs, such as Nuclear Energy Modeling and Simulation, Energy Storage Research, and the Critical Materials Institute. Funding from that department has propelled work by many scientists, including Harvard University's Pamela Silver and Daniel G. Nocera, who in 2011 made a "bionic leaf" that converted sunlight five to ten times more efficiently than plants. Their leaf was a wafer of silicon-and-solid substrates that when exposed to sunlight and water broke the water down into hydrogen and oxygen. Later work combined the bionic leaf with genetically engineered bacteria that can take hydrogen from the bionic leaf and create liquid fuel, such as isobutanol or potent ammonium nitrate fertilizer.

Science and technology can dramatically reshape the contours of prediction and risk. That's reason for hope because the historical reality is sobering. Every time Earth's climate has changed by 50 to 100 degrees Fahrenheit over the past 500 million years, things have not gone well. There have been five mass extinctions, all of them caused by changes in temperature, geochemistry, and global climate. The extinctions were devastating to life on Earth, from the Ordovician-Silurian (85 percent of species lost 440 million years ago) to the Devonian (75 percent lost 365 million years ago), the Permian-Triassic (96 percent lost 260 million years ago), the Triassic-Jurassic (80 percent lost 200 million years ago), and the Cretaceous-Paleogene (76 percent lost, including the dinosaurs, 65 million years ago). Of course, humans weren't around to initiate or remediate these extinctions. There was no species that could foresee the need to act in order to save itself, until now.

Predicting the end of the universe is far more challenging than analyzing and fixing climate change. Astronomical models depend on factors that are difficult, even impossible, to measure, such as the amount and stability of dark matter and dark energy as well as the shape and structure of a universe that will exist trillions of years in the future—well beyond the ambit of this book. In contrast, we have focused here on predictive models, data, and calculations that are trying to push out the practical prediction horizon by mere days or months or years.

The cosmologists are after bigger game.

Predictions about the demise of the universe have two major weaknesses. First, too many unknowns and deep mysteries remain. Second, the scale of space–time is bewilderingly immense, which means very small errors may lead to very large errors over long time frames. Still, over the past few centuries, our knowledge of our universe has risen exponentially. Evidence of an expanding universe goes back a century, the notion of a Big Bang to the mid-twentieth century. Stephen Hawking's work on black holes came in the 1970s; deep-space telescopes such as Hubble and Webb are even now peering more deeply into the universe's past. But scientific progress has rarely proceeded in a straight, ascending line. Will humankind continue to advance its understanding of the mechanisms of our universe? Is there a point where we can learn no more—a kind of analogous situation to perfect predictability and zero risk? Are there limitations to knowing *everything* about how this universe works? For that matter, are there other universes?

Today, astronomers predict that our solar system has a finite life span. The planets' orbits, their moons, and the trajectory of celestial bodies are, given enough time, transient and decaying.

The current science tells us that the solar system's demise will start slowly. In some 50,000 years, the length of the day will get longer due to the moon "pulling" on Earth and decelerating its rotation. Within a few million years, most of today's constellations will be unrecognizable, and Cupid and Belinda, two moons of Uranus, will likely collide, creating debris and new ring material for that planet. Current astronomical models show that the sun will increase its luminosity by 1 percent every 100 million years. By that time, most of Earth will resemble a desert. In about 180 million years, Earth's rotation will have slowed enough that a day will become 25 hours long. In 280 million years, plate tectonics will have driven the northern coast of California into Alaska. In 600 million years, the moon will have moved far enough from Earth that solar eclipses will no longer occur. In 700 million years, increasing solar luminosity will accelerate the weathering of surface rocks, and more CO_2 will be trapped as carbonate. More water will evaporate from Earth's surface, and plate tectonics will eventually slow, then stop. Most volcanic activity will cease.

Many animals will die off, and the fall in CO_2 levels will kill plants that use C3 photosynthesis—99 percent of known plant species. This will leave only C4 plants, such as maize, which can function with less water and fewer nutrients. With so much plant life eliminated, the oxygen level will begin to fall in the atmosphere, likely eliminating the ozone layer that protects most life from ultraviolet radiation.

In 800 million years, CO_2 levels will drop to a level where C4 plants can no longer survive. Without plant life cycling carbon and oxygen in the atmosphere, most multicellular life will cease to exist. In their seminal book *The Life and Death of Planet Earth* (2003), Peter Ward and Donald Brownlee project that some animal life may survive in the oceans, but even that will be difficult. Most life on the planet will not survive past 800 million years.

In a billion years, conditions will get even worse. The sun will grow about 10 percent brighter, which will drive an increase in heat on Earth. The oceans, air, and planet will absorb this heat and light, possibly triggering a greenhouse effect that could make future Earth resemble the hot hellscape of today's Venus.

The predictive task gets tougher as the time frame expands. In about 3 billion years, the sun will increase in brightness by 35 percent, which will be enough to cause the oceans to boil, whatever is left of the ice caps to melt, and significant amounts of water vapor to float into space. Life will be difficult to maintain on Earth's surface, but Mars will suddenly be more temperate. Around this time, the magnetic fields that protect Earth will likely disappear as the shifting iron inside the planet's core stops moving.

The sun will turn into a red giant in about 5 billion years. By then, everything left on Earth will be charred to a cinder. Earth's orbit around the sun will drift outward, but even with the expanded radius, the planet will likely be very close to the outer radius of the red giant sun or at the sun's surface. At that time, the planet will be reduced to a lava sphere with floating icebergs of iron and other dense matter, drifting in a bath that's 3,866 degrees Fahrenheit.

A new solar system will effectively emerge. The frigid climate of Saturn's largest moon, Titan, will become more akin to Martian conditions as they are in the twenty-first century. This means that even if we have not been able to escape the solar system in 5 billion years, we might get another few hundred million or even a billion years to work things out in the outer reaches of the solar system. But even that respite won't last forever. In about 7 billion years, Mars and Earth will become tidally locked to the sun, with one side of each planet permanently facing the light. The sun will keep growing, reaching its maximum size around 8 billion years from now, almost certainly engulfing the inner planets (Mercury, Venus, Earth, and likely Mars). This will be the sun's last big act, and if any humans have not made it out of the solar system by then, they will need to be living on the outer planets or their moons to survive.

At 10 billion years, predictions get grimmer. The sun will cool, shriveling up like a raisin into a white dwarf. The solar system will grow colder, with almost no heat and very little light. The outer planets will effectively have no light in their skies, and possibly only Mars will remain of the inner planets.

In about 100 billion years, assuming the same speed of the accelerating expansion of the universe, other galaxies will no longer be visible. This is called the "cosmic light horizon": the distances between galaxies will be so great that the time it takes for light to go from one galaxy to another will be longer than the estimated age of the universe.

In 150 billion years, the background radiation of the expanding universe will drop to about 0.3 Kelvin. In 400 billion to 500 billion years, most other stars will become invisible from any perspective. In a trillion years, new stars will cease to form. Gas clouds that now serve as star nurseries will be too far apart, and with stars hurtling away from one another, this stage will see the birth of the last star.

The end of the universe could occur in several ways. First, the expanding universe could end in a Big Freeze, an eventuality that kept Henry Adams up at night in the early twentieth century, meditating on the second law of thermodynamics. As planets drift apart, cells, molecules, atoms, and eventually even electrons and subatomic particles would be too far away from one another to interact. Unless there were a change in the universe's structure, the ever-expanding universe would get very, very cold, approaching the lowest possible temperature, 0 degrees Kelvin. Any remaining stars would run out of fuel and stop emitting energy. After trillions of years, the only remaining objects would be black holes, and even they wouldn't last forever. Even protons would decay because their half-life is predicted to be 8.2×10^{33} years. The universe would be dominated by dark matter, electrons, and positrons. All activity would cease, with any matter almost never encountering any other matter, antimatter, or anything at all. The universe would settle into random, low-energy meaninglessness—entropy, the blackest uncertainty.

There is a more hopeful alternative to death by expansion, however: the Big Crunch, in which dark matter, dark energy, and visible matter possess enough density to stop expanding and begin consolidating again. This would take a while. Over trillions of years, as matter coalesces into a singularity, a new Big Bang could occur.

Unless, of course, we're missing something. All this cosmological speculation takes us beyond science. Is there something outside the visible universe? Is there a meaning to all this—a purpose, a plan, a God? These questions verge on the philosophical or theological, just as mastering a specific prediction does not tell us much about human free will and determinism. The answers are unlikely to be revealed by predictive technologies, which are less about meaning and more about mechanism. Meanwhile, we keep acquiring new data and tools not just for the predictions we seek to make today but for those we may need in years ahead. The ever-growing Quantasaurus is a mirror of what humans have always done, applying to the future our knowledge of the past. As more data accumulate, predictions should improve, and eventually there may even be enough data to reliably predict new means of energy production or new methods of ensuring lasting peace. At least, that's the hope.

The relentless emergence of prediction in all areas of life, science, and finance will broadly reduce many forms of risk, and thus encourage some, as we've seen, to embrace greater risk. People will live longer, and they should have more productive and healthier lives. The same predictive algorithms that help us live longer and build better financial models could help us get to another world, like Mars, or to another solar system to live in a different sun's light, perhaps chauffeured by an AI-driven brain. The future is open to many possibilities, many of which we can hardly imagine, much less predict.

Acknowledgments

We thank our families, friends, and many colleagues who are mentioned in this book and are grateful for the many inspiring conversations at WorldQuant and Milken Institute events. We are grateful to Robert Teitelman, Ruth Hamel, and Michael Peltz for their steadfast and essential help on the text, ideas, and structure of the book as well as Jim Levine for helping to launch it. We thank Bob Prior for early discussions and support of the book as well as for entertaining every crazy idea we had (even if some of those ideas didn't make it into the final product). We also thank Matthew MacKay for creating many of the figures and illustrations used in the book.

Igor thanks his father for giving him encouragement and advice, Israel Englander for teaching him the asset-management business, and Stanley McChrystal for his strategic guidance. Chris thanks Joan Moriarty and Madeleine Mason-Moriarty for their fun and interesting ideas, Cory and Roseann Mason, Cory and Rebecca Mason, and Rosemary Mason for their support and edits, as well as Jeanne Maxbauer and Olivier Elemento for their leadership in the WorldQuant Initiative for Quantitative Prediction.

Bibliography

Introduction

Kay, John, and Mervyn King. *Radical Uncertainty: Decision-Making beyond the Numbers*. New York: Norton, 2020. Quotation on p. 16.

Pietruska, Jamie L. *Looking Forward: Prediction & Uncertainty in Modern America*. Chicago: University of Chicago Press, 2017. Quotations on p. 12.

Chapter 1

Döpfner, Mathias. "Interview: BioNTech Founders Özlem Türeci and Ugur Sahin on Developing the BioNTech-Pfizer COVID-19 Vaccine, the Future of Fighting Cancer, and Whether People Can Live to 200." *Business Insider*, March 23, 2021.

Garrett, Laurie. *The Coming Plague: Newly Emerging Diseases in a World out of Balance*. New York: Farrar, Straus and Giroux, 1995.

Gelles, David. "The Husband-and-Wife Team behind the Leading Vaccine to Solve Covid-19." *New York Times*, November 10, 2020.

Louten, Jennifer. "Virus Structure and Classification." *Essential Human Virology*, May 6, 2016, 19–29. https://www.ncbi.nlm.nih.gov/pmc/articles/PMC7150055/.

Oltermann, Philip. "Ugur Sahin and Özlem Türeci: German 'Dream Team' behind Vaccine." *Guardian*, November 10, 2020.

Pronker, Esther S., Tamar C. Weenen, Harry Commandeur, Eric H. J. H. M. Claassen, and Albertus D. M. E. Osterhaus. "Risk in Vaccination Research and Development Quantified." *PLOS One* 8, no. 3 (March 20, 2013). https://doi.org/10.1371/journal.pone.0057755.

Quammen, David. *Spillover: Animal Infections and the Next Human Pandemic*. New York: Norton, 2013. Quotation on p. 13.

US Centers for Disease Control and Prevention. "Ebola (Ebola Virus Disease)." Page last reviewed January 14, 2021. https://www.cdc.gov/vhf/ebola/transmission/index .html.

Waltz, Emily. "What AI Can—and Can't—Do in the Race for a Coronavirus Vaccine." *IEEE Spectrum*, September 29, 2020. https://spectrum.ieee.org/what-ai-can-and -cant-do-in-the-race-for-a-coronavirus-vaccine.

Chapter 2

Afshinnekoo, Ebrahim, Chandrima Bhattacharya, Ana Burguete-García, Eduardo Castro-Nallar, Youping Deng, Christelle Desnues, Emmanuel Dias-Neto, Eran Elhaik, Gregorio Iraola, Soojin Jang, et al.; on behalf of the MetaSUB Consortium. "COVID-19 Drug Practices Risk Antimicrobial Resistance Evolution." *Lancet Microbe* (February 24, 2021).

Bernstein, Peter L. *Against the Gods: The Remarkable Story of Risk*. New York: Wiley, 1996. Quotations on pp. 1, 336.

CDC. National Wastewater Surveillance System (NWSS). Last reviewed March 21, 2022. https://www.cdc.gov/healthywater/surveillance/wastewater-surveillance/waste water-surveillance.html.

Cornell Center for Pandemic Prevention and Response. n.d. https://pandemics.cor nell.edu/.

Danko, David, Daniela Bezdan, Evan E. Afshin, Sofia Ahsanuddin, Chandrima Bhattacharya, Daniel J. Butler, Kern Rei Chng, Daisy Donnellan, Jochen Hecht, Katelyn Jackson, et al.; International MetaSUB Consortium. "A Global Metagenomic Map of Urban Microbiomes and Antimicrobial Resistance." *Cell* 184, no. 13 (June 24, 2021): 3376–3393.

de Kraker, Marlieke, Andrew J. Stewardson, and Stephan Harbarth. "Will 10 Million People Die a Year due to Antimicrobial Resistance by 2050?" *PLOS Medicine* 13, no. 11 (November 29, 2016): e1002184.

Department of Health and Human Services (HHS). Office of the Assistant Secretary for Preparedness and Response. "Crimson Contagion 2019 Functional Exercise." October 1, 2019. https://archive.org/details/crimson-contagion-2019/mode/2up.

Knyazev, Sergey, Karishma Chhugani, Varuni Sarwal, Ram Ayyala, Harman Singh, Smruthi Karthikeyan, Dhrithi Deshpande, Pelin Icer Baykal, Zoia Comarova, Angela Lu, et al. "Unlocking Capacities of Genomics for the COVID-19 Response and Future Pandemics." *Nature Methods* 19 (April 8, 2022): 374–380.

MetaSUB International Consortium. "The Metagenomics and Metadesign of the Subways and Urban Biomes (MetaSUB) International Consortium Inaugural Meeting Report." *Microbiome* 4 (June 3, 2016): 24.

Minsky, Hyman P. "The Financial Instability Hypothesis: An Interpretation of Keynes and an Alternative to 'Standard' Theory." *Nebraska Journal of Economics and Business* 16, no. 1 (1977): 5–16.

Weill Cornell Medicine. "Dr. Jay Varma on COVID-19, Public Health Investment and the New Center for Pandemic Prevention and Response." Office of External Affairs, December 2, 2021. https://news.weill.cornell.edu/news/2021/12/dr-jay-varma -on-covid-19-public-health-investment-and-the-new-center-for-pandemic.

Chapter 3

Derman, Emanuel. "Physics Envy." Emanuelderman.com, May 16, 2002.

Dyson, George. *Darwin among the Machines: The Evolution of Global Intelligence.* New York: Basic, 1997. Quotation on p. 7.

Dyson, George. *Turing's Cathedral: The Origins of the Digital Universe.* New York: Pantheon, 2012. Quotation on p. 5.

Esteramorperez. "How to Manage Complexity and Realize the Value of Big Data." *IBM Smarter Business Review* (blog), May 28, 2020. https://www.ibm.com/blogs /services/2020/05/28/how-to-manage-complexity-and-realize-the-value-of-big-data/.

Friedman, Walter A. *Fortune Tellers: The Story of America's First Economic Forecasters.* Princeton, NJ: Princeton University Press, 2014.

Gold, E. Richard, and Julia Carbone. "Myriad Genetics: In the Eye of the Policy Storm." *Genetics in Medicine* 12 (April 1, 2010): S39–S70.

Kirkpatrick, Robert. "A New Type of Philanthropy: Donating Data." *Harvard Business Review*, March 21, 2013. https://hbr.org/2013/03/a-new-type-of-philanthropy-don.

Knight, Frank H. *Risk, Uncertainty, and Profit.* New York: Houghton Mifflin, 1921.

Li, Sheng, and Christopher E. Mason. "The Pivotal Regulatory Landscape of RNA Modifications." *Annual Review of Genomics and Human Genetics* 15 (2014): 127–150.

MacKay, Matthew, Ebrahim Afshinnekoo, Jonathan Rub, Ciaran Hassan, Mihir Khunte, Nithyashri Baskaran, Bryan Owens, Lauren Liu, Gail J. Roboz, Monica L. Guzman, et al. "The Therapeutic Landscape for Cells Engineered with Chimeric Antigen Receptors." *Nature Biotechnology* 38 (January 2020): 233–244.

Nilsson, Nils J. *The Quest for Artificial Intelligence: A History of Ideas and Achievements.* New York: Cambridge University Press, 2010. Quotation from Simon on p. 120; quotation from Minsky on p. 121.

Stephens, Zachary D., Skylar Y. Lee, Faraz Faghri, Roy H. Campbell, Chengxiang Zhai, Miles J. Efron, Ravishankar Iyer, Michael C. Schatz, Saurabh Sinha, and Gene E. Robinson. "Big Data: Astronomical or Genomical?" *PLOS Biology* 13, no. 7 (July 2015). https://doi.org/10.1371/journal.pbio.1002195.

Tulchinsky, Igor, and Robert Kirkpatrick. "The Power of Data Philanthropy." Milken Institute, June 7, 2019. https://milkeninstitute.org/articles/power-data-philanthropy.

Turing, A. M. "Computing Machinery and Intelligence." *Mind: A Quarterly Review of Psychology and Philosophy* 59, no. 236 (October 1950): 433–460.

Turing, A. M. "Lecture to the London Mathematical Society on 20 February 1947," 1–14. https://www.vordenker.de/downloads/turing-vorlesung.pdf. Quotation on p. 1.

Chapter 4

Bernstein, Peter L. *Against the Gods: The Remarkable Story of Risk*. New York: Wiley, 1996. Quotation on p. 5.

Cardano, Girolamo. *The Book on Games of Chance* (1663). Translated by Sydney Henry Gould. 1961. Reprint, New York: Holt, Rinehart and Winston, 2015.

Dreber, Anna, Coren L. Apicella, Dan T. A. Eisenberg, Justin R. Garcia, Richard S. Zamore, Koji Lum, and Benjamin Campbell. "The 7R Polymorphism in the Dopamine Receptor D4 Gene (DRD4) Is Associated with Financial Risk Taking in Men." *Evolution and Human Behavior* 30, no. 2 (March 2009): 85–92.

Dreber, Anna, David Gertler Rand, Nils Christian Wernerfelt, Justin Garcia, J. Koji Lum, and Richard J. Zeckhauser. "The Dopamine Receptor D4 Gene (DRD4) and Self-Reported Risk Taking in the Economic Domain." HKS Faculty Research Working Paper Series, John F. Kennedy School for Government, Harvard University, November 2011.

Fraga, Mario F., Esteban Ballestar, Maria F. Paz, Santiago Ropero, Fernando Setien, Maria L. Ballestar, Damia Heine-Suñer, Juan C. Cigudosa, Miguel Urioste, Javier Benitez, et al. "Epigenetic Differences Arise during the Lifetime of Monozygotic Twins." *Proceedings of the National Academy of Sciences of the United States of America* 102, no. 30 (July 26, 2005): 10604–10609.

Hacking, Ian. *The Emergence of Probability: A Philosophical Study of Early Ideas about Probability, Induction and Statistical Inference*. Cambridge: Cambridge University Press, 1975. Quotation on p. 114.

Hacking, Ian. *The Taming of Chance*. Cambridge: Cambridge University Press, 1990. Quotation from David Hume on p. 13; quotation from Pierre-Simon Laplace on pp. 11–12.

Hannum, Gregory, Justin Guinney, Ling Zhao, Li Zhang, Guy Hughes, SriniVas Sadda, Brandy Klotzle, Marina Bibikova, Jian-Bing Fan, Yuan Gao, et al. "Genome-Wide Methylation Profiles Reveal Quantitative Views of Human Aging Rates." *Molecular Cell* 49, no. 2 (January 2013): 359–367.

Hohwy, Jakob. *The Predictive Mind*. Oxford: Oxford University Press, 2013.

Horvath, Steve. "DNA Methylation Age of Human Tissues and Cell Types." *Genome Biology* 14 (December 2013). https://genomebiology.biomedcentral.com/track/pdf /10.1186/gb-2013-14-10-r115.pdf.

Kevles, Daniel J. *In the Name of Eugenics: Genetics and the Uses of Human Heredity*. New York: Knopf, 1985.

Nicolette, Timothy. "*Williams v. Quest*: The South Carolina Supreme Court's Misdiagnosis of Quest Diagnostics as a Health Care Provider and the Poor Prognosis for Plaintiffs in Medical Malpractice." *Charleston Law Review* 13, no. 3 (December 2019): 394–419.

Ray, Turna. "Quest Diagnostics Win in Wrongful Death Case Reveals Ongoing Challenges for Variant Classification." *Precision Oncology News*, November 12, 2020. https://www.360dx.com/molecular-diagnostics/quest-diagnostics-win-wrongful -death-case-reveals-ongoing-challenges-variant#.YGNDOLRKj64.

Stigler, Stephen M. *The History of Statistics: The Measurement of Uncertainty before 1900*. Cambridge, MA: Belknap Press of Harvard University Press, 1986. Quotations on pp. 63, 65.

Chapter 5

Allen, Linda J. S. "A Primer on Stochastic Epidemic Models: Formulation, Numerical Simulation, and Analysis." *Infectious Disease Modelling* 2, no. 2 (May 2017): 128–142.

Babkin, Igor V., and Irina N. Babkina. "The Origin of the Variola Virus." *Viruses* 7, no. 3 (March 2015): 1100–1112.

Bacaër, Nicolas. *A Short History of Mathematical Population Dynamics*. London: Springer, 2011.

Bernstein, Peter L. *Against the Gods: The Remarkable Story of Risk*. New York: Wiley, 1996. Quotation on p. 113; quotations from Daniel Bernoulli on p. 105.

Boyleston, Arthur. "The Origins of Inoculation." *Journal of the Royal Society of Medicine* 105, no. 7 (July 2012): 309–313.

Brambati, Bruno, and Giuseppe Simoni. "Diagnosis of Fetal Trisomy 21 in First Trimester." *Lancet*, March 12, 1983, 545–604.

Brauer, Fred. "The Kermack-McKendrick Epidemic Model Revisited." *Mathematical Biosciences* 198, no. 2 (December 2005): 119–131.

Cooper, Kelvin, Christopher Baddeley, Bernie French, Katherine Gibson, James Golden, Thiam Lee, Sadrach Pierre, Brent Weiss, and Jason Yang. "Novel Development of Predictive Feature Fingerprints to Identify Chemistry-Based Features for the Effective Drug Design of SARS-CoV-2 Target Antagonists and Inhibitors Using Machine Learning." *ACS Omega* 6, no. 7 (February 2021): 4857–4877.

Daley, D. J., and J. Gani. *Epidemic Modelling: An Introduction.* Cambridge: Cambridge University Press, 1999.

Daw, R. H. "Smallpox and the Double Decrement Table: A Piece of Actuarial Prehistory." *Journal of the Institute of Actuaries* 106, no. 3 (December 1979): 299–318.

De Vlaminck, Iwijn, Lance Martin, Michael Kertesz, Kapil Patel, Mark Kowarsky, Calvin Strehl, Garrett Cohen, Helen Luikart, Norma F. Neff, Jennifer Okamoto, et al. "Noninvasive Monitoring of Infection and Rejection after Lung Transplantation." *Proceedings of the National Academy of Sciences of the United States of America* 112, no. 43 (October 27, 2015): 13336–13341.

Dietz, Klaus, and J. A. P. Heesterbeek. "Daniel Bernoulli's Epidemiological Model Revisited." *Mathematical Biosciences* 180, nos. 1–2 (November–December 2002): 1–21.

Garrett-Bakelman, Francine E., Manjula Darshi, Stefan J. Green, Ruben C. Gur, Ling Lin, Brandon R. Macias, Miles J. McKenna, Cem Meydan, Tejaswini Mishra, Jad Masrini, et al. "The NASA Twins Study: A Multidimensional Analysis of a Year-Long Human Spaceflight." *Science* 364, no. 6436 (April 12, 2019). https://www.science.org/doi/10.1126/science.aau8650.

Hethcote, Herbert W. "The Mathematics of Infectious Diseases." *SIAM Review* 42, no. 4 (December 2000): 599–653.

Irwin, J. O. "The Place of Mathematics in Medical and Biological Statistics." *Journal of the Royal Statistical Society* 126, no. 2 (1963): 1–41.

Kermack, W. O., and A. G. McKendrick. "A Contribution to the Mathematical Theory of Epidemics." *Proceedings of the Royal Society of London* 115, no. 772 (August 1927): 700–721.

Lewontin, Richard. *It Ain't Necessarily So: The Dream of the Human Genome and Other Illusions.* New York: New York Review Books, 2000. Quotations on pp. 137–138.

Lindemann, Mary. *Medicine and Society in Early Modern Europe.* Cambridge: Cambridge University Press, 2010.

Lindqvist, Daniel, Johan Fernström, Cécile Grudet, Lennart Ljunggren, L. Träskman-Bendz, Lars Ohlsson, and Asa Westrin. "Increased Plasma Levels of Circulating

Cell-Free Mitochondrial DNA in Suicide Attempters: Associations with HPA-Axis Hyperactivity." *Translational Psychiatry* 6, no. 12 (December 2016): e971.

M'Kendrick, A. G. "Applications of Mathematics to Medical Problems." *Proceedings of the Edinburgh Mathematical Society* 44 (February 1925): 98–130. https://www .cambridge.org/core/journals/proceedings-of-the-edinburgh-mathematical-society /article/applications-of-mathematics-to-medical-problems/783296EFC2335171A044 18C1B74C248B.

PR Newswire. "New AI-Enabled Virtual Screening Tool Identifies Most Promising Approved Drugs for Success in Treating COVID-19." February 18, 2021.

Seth, Catriona. "Calculated Risks: Condorcet, Bernoulli, d'Alembert and Inoculation." *MLN* 129, no. 4 (September 2014): 740–755. Quotation on pp. 747–748.

Topol, Eric J. "Individualized Medicine from Prewomb to Tomb." *Cell* 157, no. 1 (March 27, 2014): 241–253.

Weill Cornell Medicine. "Mapping Cancer's Drug Resistance to Design Better Treatment Regimens." October 23, 2019.

Williams, Gareth. *Angel of Death: The Story of Smallpox.* New York: Palgrave Macmillan, 2010.

Chapter 6

Arrow, Kenneth J. "Uncertainty and the Welfare Economics of Medical Care." *American Economic Review* 53, no. 5 (December 1963): 941–973. Quotations on p. 961.

Baker, Tom. "On the Genealogy of Moral Hazard." *Texas Law Review* 75, no. 2 (December 1996): 237–292. Quotation on p. 255.

Balasubramanian, Ramnath, Ari Libarikian, and Doug McElhany. "Insurance 2030— the Impact of AI on the Future of Insurance." McKinsey & Co., March 2021. https:// www.mckinsey.com/industries/financial-services/our-insights/insurance-2030-the -impact-of-ai-on-the-future-of-insurance.

Bellhouse, David R. "A New Look at Halley's Life Table." *Journal of the Royal Statistical Society* 174, no. 3 (July 2011): 823–832.

Brewer, John. *Sinews of Power: War, Money and the English State, 1688–1783.* Cambridge, MA: Harvard University Press, 1990. Quotations on p. 74.

Falchuk, Bryan. *The Future of Insurance: From Disruption to Evolution.* St. Paul, MN: Insurance Nerds, 2020.

Frier, Bruce. "Roman Life Expectancy: Ulpian's Evidence." *Harvard Studies in Classical Philology* 86 (1982): 213–251. Quotations and statistics on p. 213.

Gelderblom, Oscar, and Joost Jonker. "Public Finance and Economic Growth: The Case of Holland in the Seventeenth Century." *Journal of Economic History* 71, no. 1 (March 2011): 1–39.

Goetzmann, William N. *Money Changes Everything: How Finance Made Civilization Possible.* Princeton, NJ: Princeton University Press, 2016. Quotations on pp. 256, 258.

Hacking, Ian. *The Emergence of Probability: A Philosophical Study of Early Ideas about Probability, Induction and Statistical Inference.* Cambridge: Cambridge University Press, 1975. Quotation on p. 113.

Hup, Mark. "Life Annuities as a Resource of Public Finance in Holland, 1648–1713: Demand or Supply-Driven?" Netspar, July 2011. https://www.netspar.nl/assets /uploads/038_MSc_Mark_Hup.pdf.

Pauly, Mark V. "The Economics of Moral Hazard: Comment." *American Economic Review* 58, no. 3 (June 1968): 531–537. Quotations on p. 537.

Powers, Michael R. *Acts of God and Man: Ruminations on Risk and Insurance.* New York: Columbia Business School Publishing, 2011. Quotations on pp. 3, 7, 13.

Price, Richard. *Observations on Reversionary Payments; on Schemes for Providing Annuities for Widows, and for Persons in Old Age; on the Method of Calculating the Values of Assurances on Lives; and on the National Debt. To Which Are Added, Four Essays on Different Subjects in the Doctrine of Life-Annuities and Political Arithmetick.* London: T. Cadell, 1773.

Rowell, David, and Luke B. Connelly. "A History of the Term 'Moral Hazard.'" *Journal of Risk and Insurance* 79, no. 4 (December 2012): 1051–1075. Quotations on p. 1058.

Syme, Ronald. "Lawyers in Government: The Case of Ulpian." *Proceedings of the American Philosophical Society* 116, no. 5 (October 1972): 406–409.

Turnbull, Craig. *A History of British Actuarial Thought.* London: Palgrave Macmillan, 2017. Quotations on p. 64.

Woods, Robert. "Ancient and Early Modern Mortality: Experience and Understanding." *Economic History Review* 60, no. 2 (May 2007): 373–399.

Chapter 7

Afshinnekoo, Ebrahim, Cem Meydan, Shanin Chowdhury, Dyala Jaroudi, Collin Boyer, Nick Bernstein, Julia M. Maritz, Darryl Reeves, Jorge Gandara, Sagar Chhangawala, et al. "Geospatial Resolution of Human and Bacterial Diversity with City-Scale Metagenomics." *Cell Systems* 1, no. 1 (July 2015): 72–87.

Ehrlich, Yaniv. "A Vision for Ubiquitous Sequencing." *Genome Research* 25, no. 10 (October 2015): 1411–1416.

Gilbert, Natasha. "Why the 'Devious Defecator' Case Is a Landmark for US Genetic-Privacy Law." *Nature*, June 25, 2015. https://www.nature.com/articles/nature.2015 .17857.

Joly, Yann, Charles Dupras, Miriam Pinkesz, Stacey A. Tovino, and Mark A. Roth-stein. "Looking beyond GINA: Policy Approaches to Address Genetic Discrimina-tion." *Annual Review of Genomics and Human Genetics* 21 (August 2020): 491–507. https://www.annualreviews.org/doi/abs/10.1146/annurev-genom-111119-011436.

Krishna, Dean. "DNA Testing for Eddy Curry? Creating a New Constitutional Protec-tion." *University of Pennsylvania Journal of Constitutional Law* 9, no. 4 (April 2007): 1105–1129.

Kyllo v. United States. 533 US 27 (2001). https://supreme.justia.com/cases/federal /us/533/27/.

Larsen, David A., and Krista R. Wigginton. "Tracking COVID-19 with Wastewater." *Nature Biotechnology* 38 (September 2020): 1151–1153.

Massachusetts Institute of Technology Underworlds Project. N.d. http://under worlds.mit.edu/.

Menge, Susan K. "Should Players Have to Pass to Play? A Legal Analysis of Imple-menting Genetic Testing in the National Basketball Association." *Marquette Sports Law Review* 17, no. 2 (Spring 2007): 459–480.

Mittelman, David. Interview with the authors. June 24, 2021.

Pettigrew, Matthew. "Smart-Toilets & Biometrics." *Predict*, October 26, 2020. https://medium.com/predict/smart-toilets-biometrics-7f1070a70a05.

Philipkoski, Kristen. "Genetic Testing Case Settled." *Wired*, April 10, 2001.

Rawls, John. *A Theory of Justice.* Cambridge, MA: Belknap Press of Harvard University Press, 1971.

Robertson, Adi. "How Target Can Detect Pregnancies, Create New Habits, and Know What You Want to Do Before You Do." *Verge*, February 17, 2012. https://www.the verge.com/2012/2/17/2804554/target-data-mining-advertising-pregnancy-predic tion.

Rosenfeld, Jeffrey A., and Christopher E. Mason. "Pervasive Sequence Patents Cover the Entire Human Genome." *Genome Medicine* 5, no. 3 (March 2013): 27.

Uberti, David. "After Backlash, Predictive Policing Adapts to a Changed World." *Wall Street Journal*, July 8, 2021.

Chapter 8

Associated Press. "Pentagon to Build 'Safer' Cluster Bombs." July 7, 2008. https://www.nbcnews.com/id/wbna25576982.

Awad, Edmond, Sohan Dsouza, Richard Kim, Jonathan Schulz, Joseph Henrich, Azim Shariff, Jean-François Bonnefon, and Iyad Rahwan. "The Moral Machine Experiment." *Nature* 563, no. 7729 (November 2018): 59–64. Quotation on p. 62.

Bostrom, Nick, and Eliezer Yudkowsky. "The Ethics of Artificial Intelligence." Machine Intelligence Research Institute, 2011. https://intelligence.org/files/EthicsofAI.pdf.

Bureau of Investigative Journalism. *2019 Annual Report.* London: Bureau of Investigative Journalism, 2019.

Chavannes, Esther, and Amit Arkhipov-Goyal. "Towards Responsible Autonomy: The Ethics of RAS in a Military Context." Hague Centre for Strategic Studies, September 12, 2019.

Clifford, Catherine. "Elon Musk: 'Robots Will Be Able to Do Everything Better than Us.'" CNBC, July 17, 2017. http://www.cnbc.com/2017/07/17/elon-musk-robots-will-be-able-to-do-everything-better-than-us.html.

Horowitz, Michael C. "The Ethics & Morality of Robotic Warfare: Assessing the Debate over Autonomous Weapons." *Daedalus* 145, no. 4 (Fall 2016): 25–36. Quotation on p. 26.

Lal, Sangeeta, Lipika Tiwari, Ravi Ranjan, Ayushi Verma, Neetu Sardana, and Rahul Mourya. "Analysis and Classification of Crime Tweets." *Procedia Computer Science* 167 (2020): 1911–1919.

Metz, Cade. "Police Drones Are Starting to Think for Themselves." *New York Times,* December 5, 2020.

Pegasystems. "What Consumers Really Think about AI: A Global Study." August 15, 2017. https://www.ciosummits.com/what-consumers-really-think-about-ai.pdf.

Pinker, Steven. *The Better Angels of Our Nature: Why Violence Has Declined.* New York: Viking, 2011.

Porter, Tom. "AI Could Spark Wars Unless Legislators Act Now, Warns Elon Musk." *Newsweek,* July 16, 2017.

Rogin, Josh. "The Trump Administration Cancels a Plan to Curtail the Use of Cluster Bombs." *Washington Post,* November 30, 2017. Quotation from the Trump administration.

Scharre, Paul. *Army of None: Autonomous Weapons and the Future of War.* New York: Norton, 2018. Quotations on pp. 5, 12, 21, 42.

Tegmark, Max. *Life 3.0: Being Human in the Age of Artificial Intelligence.* New York: Knopf, 2017.

Tegmark, Max, and Stuart Russell. "Autonomous Weapons: An Open Letter from AI & Robotics Researchers." Future of Life Institute, February 9, 2016.

Ticehurst, Rupert. "The Martens Clause and the Laws of Armed Conflict." *International Review of the Red Cross*, no. 317 (April 1997): 125–134.

US Department of Defense. "DOD Adopts Ethical Principles for Artificial Intelligence." February 24, 2020. https://www.defense.gov/Newsroom/Releases/Release /Article/2091996/dod-adopts-ethical-principles-for-artificial-intelligence/.

US Department of Defense. "DoD Policy on Cluster Munitions." November 30, 2017. https://dod.defense.gov/Portals/1/Documents/pubs/DOD-POLICY-ON-CLUSTER -MUNITIONS-OSD071415-17.pdf.

Chapter 9

Black, J. Stewart, and Patrick van Esch. "AI-Enabled Recruiting: What Is It and How Should a Manager Use It?" *Business Horizons* 63, no. 2 (March–April 2020): 215–226. Quotations on pp. 216, 219.

Boomsma, Dorret, Andreas Busjahn, and Leena Peltonen. "Classical Twin Studies and Beyond." *Nature Reviews Genetics* 3 (November 2002): 872–882.

Capelli, Peter. "Your Approach to Hiring Is All Wrong." *Harvard Business Review*, May–June 2019. Quotations on pp. 1–3; quotation from Korn Ferry on p. 3. https://hbr.org /2019/05/your-approach-to-hiring-is-all-wrong.

Dattner, Ben, Tomas Chamorro-Premuzic, Richard Buchband, and Lucinda Schettler. "The Legal and Ethical Implications of Using AI in Hiring." *Harvard Business Review*, April 25, 2019. https://hbr.org/2019/04/the-legal-and-ethical-implications -of-using-ai-in-hiring.

Hall, Joshua D., Anna B. O'Connell, and Jeanette G. Cook. "Predictors of Student Productivity in Biomedical Graduate School Admissions." *PLOS One* 12, no. 1 (January 11, 2017). https://psycnet.apa.org/record/2017-03921-001.

Harvard Molecular Technologies. "Protective Alleles." Table, n.d. https://arep.med .harvard.edu/gmc/protect.html.

Holtzman, Liad, and Charles A. Gersbach. "Editing the Epigenome: Reshaping the Genomic Landscape." *Annual Review of Genomics and Human Genetics* 19 (August 2018): 43–71.

Kramer, Miriam, and Bryan Walsh. "Radiation-Proofing the Human Body for Long-Term Space Travel." Axios, September 29, 2020. https://www.axios.com/gene-editing -radiation-space-travel-af13be1e-c202-4020-a491-d4d9b6d5593c.html.

Liu, X. Shawn, Hao Wu, Xiong Ji, Yonatan Stelzer, Xuebing Wu, Szymon Czauderna, Jian Shu, Daniel Dadon, Richard A. Young, and Rudolf Jaenisch. "Editing DNA Methylation in the Mammalian Genome." Cell 167, no. 1 (September 2016): 233–247.

Liu, X. Shawn, Hao Wu, Marine Krzisch, Xuebing Wu, John Graef, Julien Muffat, Denes Hnisz, Charles H. Li, Bingbing Yuan, Chuanyun Xu, et al. "Rescue of Fragile X Syndrome Neurons by DNA Methylation Editing of the FMR1 Gene." Cell 172, no. 5 (February 2018): 979–992. https://pubmed.ncbi.nlm.nih.gov/29456084/.

Moneta-Koehler, Liane, Abigail M. Brown, Kimberly A. Petrie, Brent J. Evans, and Roger Chalkley. "The Limitations of the GRE in Predicting Success in Biomedical Graduate School." PLOS One 12, no. 1 (January 2017). https://journals.plos.org/ plosone/article?id=10.1371/journal.pone.0166742.

Mullin, Gemma. "Astronaut Scott Kelly's DNA No Longer Matches Identical Twin's after a Year in Space, NASA Says." Sun, March 15, 2018.

Chapter 10

Campbell, W. Joseph. Lost in a Gallup: Polling Failure in U.S. Presidential Elections. Oakland: University of California Press, 2020.

Fukuyama, Francis. The End of History and the Last Man. New York: Free Press, 1992.

Greenberg, Daniel S. "The Plague of Polling." Baltimore Sun, June 25, 1996.

Halpern, Sue. "Cambridge Analytica and the Perils of Psychographics." New Yorker, March 30, 2018.

Hern, Alex. "Cambridge Analytica: How Did It Turn Clicks into Votes?" Guardian, May 6, 2018.

Jackson, Natalie. "Poll-Based Election Forecasts Will Always Struggle with Uncertainty." Sabato's Crystal Ball, August 6, 2020. https://centerforpolitics.org/crystalball /articles/poll-based-election-forecasts-will-always-struggle-with-uncertainty/.

Jackson, Natalie. "Trump–Biden 2020: The Polls Were Off, but They're Not Crystal Balls and Aren't Meant to Be." USA Today, December 5, 2020.

Keeter, Scott, Courtney Kennedy, and Claudia Deane. "Understanding How 2020 Election Polls Performed and What It Might Mean for Other Kinds of Survey Work." Pew Research Center, November 13, 2020.

Kramer, Adam D. I., Jamie E. Guillory, and Jeffrey T. Hancock. "Experimental Evidence of Massive-Scale Emotional Contagion through Social Networks." *Proceedings of the National Academy of Sciences of the United States of America* 111, no. 24 (June 2014): 8788–8790. Quotations on p. 8790.

Lippmann, Walter. *Public Opinion*. New York: Harcourt, Brace, 1922.

Silver, Nate. "The Polls Weren't Great. But That's Pretty Normal." FiveThirtyEight, November 11, 2020. https://fivethirtyeight.com//features/the-polls-werent-great-but-thats-pretty-normal/.

Sumner, Chris. "Rage Against the Weaponized AI Propaganda Machine." Def Con 25, October 20, 2017. https://www.youtube.com/watch?v=g5Hx86H3-mc.

Sumner, Chris, John E. Scofield, Erin Buchanan, Mimi-Rose Evans, and Matthew Shearing. "The Role of Personality, Authoritarianism and Cognition in the United Kingdom's 2016 Referendum on European Union Membership." ResearchGate, July 2018. https://www.researchgate.net/publication/326163776_The_Role_of_Personality _Authoritarianism_and_Cognition_in_the_United_Kingdom's_2016_Referendum_on _European_Union_Membership.

Warren, Kenneth F. *In Defense of Public Opinion Polling*. Boulder, CO: Westview Press, 2001.

Wong, Julia Carrie, "Cambridge Analytica-Linked Academic Spurns Idea Facebook Swayed Election." *Guardian*, June 19, 2018. Quotation from Aleksandr Kogan during congressional testimony.

Wylie, Christopher. *Mindf*ck: Inside Cambridge Analytica's Plot to Break the World*. New York: Random House, 2019. Quotations on pp. 82, 91, 92.

Zitner, Aaron. "Trump–Biden Was Worst Presidential Polling Miss in 40 years, Panel Says." *Wall Street Journal*, May 13, 2021.

Zuboff, Shoshana. *The Age of Surveillance Capitalism: The Fight for a Human Future at the New Frontier of Power*. New York: PublicAffairs, 2019.

Zuboff, Shoshana. "The Coup We Are Not Talking About." *New York Times*, January 29, 2021. https://www.nytimes.com/2021/01/29/opinion/sunday/facebook -surveillance-society-technology.html.

Chapter 11

Broadberry, Stephen, Bruce M. S. Campbell, Alexander Klein, Mark Overton, and Bas van Leeuwen. *British Economic Growth: 1270–1870*. Cambridge: Cambridge University Press, 2015.

Hacking, Ian. *The Taming of Chance*. Cambridge: Cambridge University Press, 1990. Quotations on p. 200.

Hum, Derek, and Wayne Simpson. "Economic Response to a Guaranteed Annual Income: Experience from Canada and the United States." *Journal of Labor Economics* 11, no. 1, part 2 (January 1993): S263–S296.

Karpathy, Andrej. "The Unreasonable Effectiveness of Recurrent Neural Networks." Andrej Karpathy blog, May 21, 2015. http://karpathy.github.io/2015/05/21/rnn-effectiveness.

Knight, Frank H. *Risk, Uncertainty, and Profit*. New York: Houghton Mifflin, 1921.

Menand, Louis. *The Metaphysical Club: A Story of Ideas in America*. New York: Farrar, Straus and Giroux, 2001. Quotation on pp. 199–200.

Rifkin, Jeremy. *The End of Work: The Decline of the Global Labor Force and the Dawn of the Post-market Era*. New York: Jeremy P. Tarcher and Penguin, 1996. Quotations on p. 192.

Schumpeter, Joseph A. *Capitalism, Socialism, and Democracy*. New York: Harper, 1942. Quotations on pp. 77, 83.

Simpson, Wayne, Greg Mason, and Ryan Godwin. "The Manitoba Basic Annual Income Experiment: Lessons Learned 40 Years Later." *Canadian Public Policy* 43, no. 1 (March 2017): 85–104.

UBS and PwC Switzerland. *Riding the Storm: Market Turbulence Accelerates Diverging Fortunes*. Geneva: UBS and PwC Switzerland, 2020. https://www.ubs.com/content/dam/static/noindex/wealth-management/ubs-billionaires-report-2020-spread.pdf.

US Centers for Disease Control and Prevention. "Trends in the Use of Telehealth During the Emergence of the COVID-19 Pandemic—United States, January–March 2020." October 30, 2020. https://www.cdc.gov/mmwr/volumes/69/wr/mm6943a3.htm.

Afterword

Clark, Peter U., Jeremy D. Shakun, Shaun A. Marcott, Alan C. Mix, Michael Eby, Scott Kulp, Anders Levermann, Glenn A. Milne, Patrik L. Pfister, Benjamin D. Santer, et al. "Consequences of Twenty-First-Century Policy for Multi-millennial Climate and Sea-Level Change." *Nature Climate Change* 6, no. 4 (February 2016): 360–369.

Torella, Joseph P., Christopher J. Gagliardi, Janice S. Chen, D. Kwabena Bediako, Brendan Colón, Jeffery C. Way, Pamela A. Silver, and Daniel G. Nocera. "Efficient Solar-to-Fuels Production from a Hybrid Microbial-Water-Splitting Catalyst System." *Proceedings of the National Academy of Sciences of the United States of America* 112, no. 8 (February 2015): 2337–2342.

Voosen, Paul. "A 500-Million-Year Survey of Earth's Climate Reveals Dire Warning for Humanity." *Science*, May 22, 2019. https://www.science.org/content/article /500-million-year-survey-earths-climate-reveals-dire-warning-humanity.

Ward, Peter D., and Donald Brownlee. *The Life and Death of Planet Earth: How the New Science of Astrobiology Charts the Ultimate Fate of Our World*. New York: Times Books, 2003.

Wilder, Rob, and Dan Kammen. "Taking the Long View: The 'Forever Legacy' of Climate Change." *Yale Environment 360*, September 12, 2017.

Index